Liquid Crystals and Other Partially Disordered Molecular Systems

Liquid Crystals and Other Partially Disordered Molecular Systems

Guest Editor

Dorota Dardas

MDPI

Basel • Beijing • Wuhan • Barcelona • Belgrade • Novi Sad • Cluj • Manchester

Guest Editor
Dorota Dardas
Institute of Molecular Physics
Polish Academy of Sciences
Poznań
Poland

Editorial Office
MDPI AG
Grosspeteranlage 5
4052 Basel, Switzerland

This is a reprint of the Special Issue, published open access by the journal *Materials* (ISSN 1996-1944), freely accessible at: https://www.mdpi.com/journal/materials/special_issues/BS138SG582.

For citation purposes, cite each article independently as indicated on the article page online and as indicated below:

Lastname, A.A.; Lastname, B.B. Article Title. *Journal Name* **Year**, *Volume Number*, Page Range.

ISBN 978-3-7258-6762-2 (Hbk)
ISBN 978-3-7258-6763-9 (PDF)
https://doi.org/10.3390/books978-3-7258-6763-9

Cover image courtesy of Dorota Dardas

Contents

About the Editor

Dorota Dardas

Dorota Dardas is a researcher at the Institute of Molecular Physics of the Polish Academy of Sciences. Her scientific interests focus on the physical properties of liquid crystal materials, in particular ferroelectric and antiferroelectric phases, as well as their viscoelastic and dielectric behavior. She has participated in numerous national and international projects devoted to the design, characterization, and application of functional soft matter systems. Her recent research includes a comprehensive analysis of viscoelastic effects and dielectric modes in chiral liquid crystals, contributing to a deeper understanding of their dynamic processes and potential technological applications.

In addition to her scientific research, Dorota Dardas is actively involved in science communication, education, and social activities. In the Institute of Molecular Physics, Polish Academy of Sciences, she coordinates the 'Researchers' Night' projects and promotes inclusive education aimed at broadening access to scientific knowledge. She is an ambassador for the TYGIEL Foundation, supports interdisciplinary cooperation, and participates in the organizational and scientific committees of national and international conferences. She also cooperates with businesses and social organizations to support cooperation between science and society. In addition, she is an active member of the Relatives Foundation and co-founder of the magazine 'Człowiek i Relacje' (Man and Relationships), which promotes dialogue between science, education, and social engagement.

Preface

The Special Issue "Liquid Crystals and Other Partially Disordered Molecular Systems" presents research on the structure, dynamics, and properties of materials with partial molecular order, aimed at deepening our understanding of phase behaviour and functionality in complex soft and hybrid systems.

As the Guest Editor, I would like to thank *Materials* for its support of this Reprint and acknowledge the Editorial Board Members of this Reprint and Ms. Jenna Peng for their invaluable assistance in organizing and publishing this issue. I appreciate all the authors for their pioneering research and exploration into frontier issues within this field.

Dorota Dardas
Guest Editor

materials

MDPI

Article

Plasmid DNA Complexes in Powder Form Studied by Spectroscopic and Diffraction Methods

Aleksandra Radko [1], Sebastian Lalik [1], Natalia Górska [2], Aleksandra Deptuch [3], Jolanta Świergiel [4] and Monika Marzec [1,*]

1 Institute of Physics, Jagiellonian University, Łojasiewicza 11, 30-348 Kraków, Poland; sebastian.lalik@uj.edu.pl (S.L.)

2 Faculty of Chemistry, Jagiellonian University, Gronostajowa 2, 30-387 Kraków, Poland; natalia.gorska@uj.edu.pl

3 Institute of Nuclear Physics Polish Academy of Sciences, Radzikowskiego 152, 31-342 Kraków, Poland; aleksandra.deptuch@ifj.edu.pl

4 Institute of Molecular Physics Polish Academy of Sciences, Smoluchowskiego 17, 60-179 Poznań, Poland; swiergiel@ifmpan.poznan.pl

* Correspondence: monika.marzec@uj.edu.pl

Abstract: Currently, new functional materials are being created with a strong emphasis on their ecological aspect. Materials and devices based on DNA biopolymers, being environmentally friendly, are therefore very interesting from the point of view of applications. In this paper, we present the results of research on complexes in the powder form based on plasmid DNA (pDNA) and three surfactants with aliphatic chains containing 16 carbon atoms (cetyltrimethylammonium chloride, benzyldimethylhexadecylammonium chloride and hexadecylpyridinium chloride). The X-ray diffraction results indicate a local hexagonal packing of DNA helices in plasmid DNA complexes, resembling the packing for corresponding complexes based on linear DNA. Based on the Fourier-transform infrared spectroscopy results, the DNA conformation in all three complexes was determined as predominantly of A-type. The two relaxation processes revealed by dielectric spectroscopy for all the studied complexes are connected with two different contributions to total conductivity (crystallite part and grain boundaries). The crystallite part (grain interior) was interpreted as an oscillation of the polar surfactant head groups and is dependent on the conformation of the surfactant chain. The influence of the DNA type on the properties of the complexes is discussed, taking into account our previous results for complexes based on linear DNA. We showed that the type of DNA has an impact on the properties of the complexes, which has not been demonstrated so far. It was also found that the layer of pDNA–surfactant complexes can be used as a layer with variable specific electric conductivity by selecting the frequency, which is interesting from an application point of view.

Keywords: plasmid DNA; DNA complex; dielectric spectroscopy; X-ray diffraction; infrared spectroscopy

1. Introduction

The research on deoxyribonucleic acid (DNA) was inextricably linked to and motivated mainly by genetics, while the discovery of other, potentially useful properties of DNA began only in the 21st century. From a chemical point of view, deoxyribonucleic acid belongs to the group of macromolecular organic compounds, the group of nucleic acids. DNA is a linear, unbranched biopolymer, and its monomer consists of 2-deoxyribose ($C_5H_{10}O_4$), which connects to the phosphate residue PO_4 through the 5′ carbon atom [1]. On the other hand, the C1 carbon atom forms an N-glycosidic bond with one of the nitrogenous bases (adenine A, guanine G, thymine T, cytosine C) [2,3]. Linear double-stranded DNA (dsDNA) is the most studied form, but there are also other linear forms: single-stranded (ssDNA), triple-stranded and four-stranded. The double-stranded DNA can take various forms depending on environment (e.g., the humidity, pH, chemical environment or salt content). The most common are two right-handed forms (A-DNA and B-DNA) and one left-handed form (Z-DNA). The main form is B-DNA; the A-DNA form is formed at reduced humidity from the B-DNA form, while the Z-DNA form occurs in an environment with a high salt content. Additionally, DNA can also exist in the form of small circular molecules called plasmids. Plasmid molecules are usually smaller than chromosomal DNA molecules and consist of 300 bp to 2400 kbp (human chromosomal DNA has a length of approx. 4700 kbp—approx. 250 Mbp) [4]. Plasmid DNA can occur in three forms: circular, linear and supercoiled [5].

It should be also noted here that the chemical bonds that make up the individual strands of DNA and the double helix as a whole are weak (e.g., hydrogen bonds between base pairs), but the molecule as a whole is very stable [6]. The helical form of DNA contains large and small grooves in the outer surface of the molecule. In DNA solutions, there are bound water molecules in these grooves. Some cations (counterions) bind to the phosphate backbone with a weak covalent bond. On the other hand, other cations are loosely bound and can penetrate into the grooves of the DNA helix. The counterions bind close to the negative charge centres of the DNA, and a double layer is formed. Ions attracted to the charged DNA backbone form a shell that shields some of the negative charge of the DNA. The dominant charges on the DNA molecule come from the oxygen atoms that are part of the phosphate groups located on the outer surfaces of the double helix. In solutions, the shell around the DNA double helix is usually formed by Na^+ or Mg^{2+} counterions, which are attracted to the backbone with negatively charged phosphate groups. These counterions have some mobility and oscillate around the phosphate charge centres in an external alternating electric field (measuring field). However, when a cationic surfactant is used, the counterions are surfactant cations, for example, for the surfactant cetyltrimethylammonium chloride (CTMA), it is $CTMA^+$.

It is commonly believed that a double-stranded DNA molecule has a permanent dipole moment [6]. The double-stranded DNA macromolecule is negatively charged along the entire length of the double helix due to the presence of phosphate residues. The dipole moment of one strand is opposite to the dipole moment of the other strand. Thus, the net dipole moment of double-stranded DNA should be zero. When DNA is dissolved in a cationic solution, a part of the negative charges is neutralized by the cations. Therefore, the addition of cationic surfactants to the phosphate backbone of the DNA double helix causes symmetry breaking, and a dipole moment is induced. Moreover, the deformation of the DNA molecule by the external alternating electric field applied may cause a non-zero net dipole moment of double-stranded DNA without cationic surfactants.

The very well-known properties of the DNA biomacromolecule cause great interest in this material in various fields of science [7–11]. In recent decades, many applications have been found for it outside of biological sciences, most often in the form of DNA–cationic surfactant complex. Such complexes are characterized by a very high application potential with the possibility of implementation in various fields, from medicine through photonics to organic electronics. Most commonly, such compounds are used in life science to act as drug delivery systems [12]. Outside life science, DNA complexes are currently being used to create new functional materials for applications in, for example, organic thin-film transistors and biosensors [13–17], as well as being a basic building block in the field of supramolecular chemistry for conducting molecular calculations [18,19].

The study of the properties of DNA complexes is therefore very interesting from the point of view of their applications, also considering the ecological aspect of the new functional materials and devices based on them. It is known that complexation with cationic surfactants improves the optical properties of DNA [11]. The question therefore arises whether such a cationic surfactant will change/improve the dielectric properties of the material. There are also many articles that do not specify a DNA type. Our idea is to check whether the type of DNA influences the properties of the complexes. Hence, we synthesized complexes based on linear and plasmid DNA types with three selected surfactants. Moreover, we conducted this research on samples in the form of powder, not solutions. This type of research is not common and that is why we decided to conduct it. The properties of DNA–surfactant complexes in the form of powders based on linear DNA have already been studied by us [20]. In this paper, we describe the properties of DNA–surfactant complexes based on plasmid DNA, compare them with the properties of complexes based on linear DNA and show that the type of DNA has an impact on the properties of the complexes, which has not been demonstrated so far. We investigated the influence of DNA type and surfactant type on the permittivity, specific electric conductivity, relaxation processes and average distances between the aliphatic surfactant chains in the resulting complexes. The research was carried out by means of Fourier-transform middle-infrared (FT-MIR) and frequency-domain dielectric spectroscopy (FDDS) as well as X-ray diffraction (XRD). FT-MIR spectroscopy was used to check whether complexes were formed and what DNA conformation exists in these complexes, while XRD studies were used to find a local packing of DNA helices in the complexes synthesized. In turn, the FDDS method is widely used because it is very sensitive, and the result is obtained quickly. In this method, the sample's reaction to an applied weak alternating electric field (the so-called measuring field) is recorded. As mentioned above, DNA–surfactant complexes exhibit a dipole moment, and the interaction between the measuring field and the dipole moment can be recorded in FDDS, so this method is useful. In our research, it was used to study the electric properties of the synthesized complexes. The electric properties of the complexes are important because, as previously mentioned, the possibilities of their use in organic thin-film transistors or biosensors are already being investigated.

2. Materials and Methods

2.1. Materials

Plasmid DNA, phMGFP (Promega, Madison, WI, USA), was isolated from transformed Escherichia coli bacteria with a NucleoBond 10 000 EF DNA extraction kit (Macherey-Nagel, Düren, Germany) as described previously [21]. Cationic surfactants: cetyltrimethylammonium chloride (CTMA, $CH_3(CH_2)_{15}N(Cl)(CH_3)_3$), hexadecylpyridinium chloride (HDP, $C_{21}H_{38}ClN{\cdot}H_2O$) and benzyldimethylhexadecyl-ammonium chloride (BAC, $CH_3(CH_2)_{15}N(Cl)(CH_3)_2CH_2C_6H_5$) as well as Tris base (tris(hydroxymethyl) aminomethane) and hydrochloric acid (HCl) were used as received from Sigma-Aldrich (St. Louis, MO, USA).

The applied cationic surfactants BAC, CTMA and HDP have in their molecular structure an aliphatic chain containing 16 $-CH_2-$ groups ($-(CH_2)_{15}CH_3$); the difference is only in the end of this chain. The BAC surfactant has the highest molecular weight, while the CTMA surfactant has the lowest. For each surfactant, the positive charge is deposited on the nitrogen atom, which, in each case, has a completely different chemical surrounding. It is through this nitrogen atom that the surfactant interacts with the negatively charged double helix of pDNA (or rather, the unsaturated oxygen atoms of the phosphate residues), resulting in a charged double layer and the formation of the complex. The pDNA–surfactant complexes were synthesized according to the procedure described by us in [20].

2.2. Methods

Fourier-transform middle-infrared absorption measurements (FT-MIR) were performed at room temperature in transmission mode using a Bruker VERTEX 70v vacuum spectrometer (Billerica, MA, USA). The spectra were obtained in the range of 400–4000 cm^{-1} with a spectral resolution of 2 cm^{-1} and 32 scans per spectrum. Bulk samples of the pure components (dried pDNA, BAC, CTMA and HDP surfactant) as well as the pDNA–surfactant complexes were mixed with KBr, compressed into pellets and measured at room temperature.

X-ray diffraction (XRD) measurements for polycrystalline pDNA complexes (in borosilicate glass capillaries with a 0.5 mm outer diameter) were carried out in the $2\theta = 2–40°$ range with an Empyrean 2 (PANalytical, Malvern, UK) diffractometer (CuKα radiation, parabolic mirror in the incident beam). The diffraction pattern of an empty capillary was used for background subtraction. The results were analysed using WinPLOTR [22] and Origin.

Frequency domain dielectric spectroscopy (FDDS) measurements were performed by using the Turnkey Impedance Spectrometer Concept 81 (Novocontrol Technologies GmbH & Co. KG, Montabaur, Germany) at room temperature (23.0 ± 0.5 °C) in the frequency range of 0.5 Hz–3.0 MHz, at a measuring voltage of 0.1 V. All pellet samples were placed in a flat, rounded electrode capacitor (diameter of 9.60 mm). The thickness of each sample was determined by a micrometric screw (see Table 1).

Table 1. The thickness of the studied pDNA–surfactant complex.

Sample	Thickness [mm]
pDNA–BAC	0.30(1)
pDNA–HDP	0.35(1)
pDNA–CTMA	0.35(1)

3. Results and Discussion

The FT-MIR spectra of the pure components: plasmid DNA, CTMA, HDP and BAC surfactants as well as the synthesized plasmid DNA–surfactant complexes are presented in Figure 1. In the spectrum of pure plasmid DNA, the bands at 832, 962, 1290 and 1418 cm^{-1} characteristic for the B-conformation of DNA are present. In addition to them, the bands that are spectroscopic markers for the A-DNA form are observed at 801, 860, 1176 and 1238 cm^{-1}. There is also one band visible at 1208 cm^{-1}, which, in turn, can be associated with the Z-DNA form. Thus, the pure plasmid DNA substrate used in our study was most probably a mixture of the three conformations, with the B and A forms being dominant. The molecular structure of pDNA differs from that observed for linear DNA substrate used in preparation of the analogical DNA–surfactant complexes presented in our previous work, in which the B-DNA form was predominant [20]. A comparison of the IR spectra of bulk plasmid and linear DNA along with a description of their bands is presented in the Appendix A.

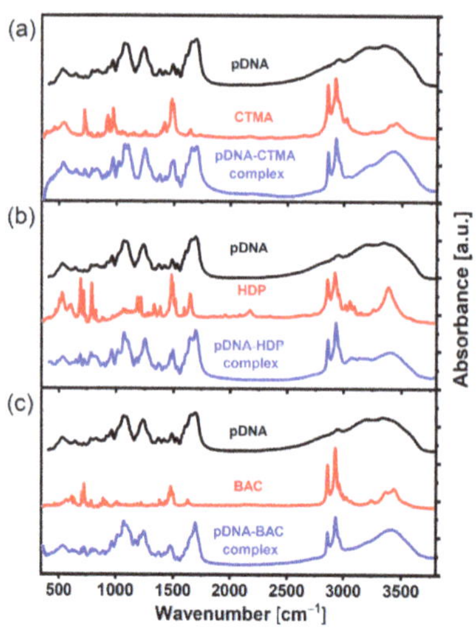

Figure 1. FT-MIR spectra of pure components (pDNA, CTMA, HDP and BAC) and obtained complexes: pDNA–CTMA (**a**), pDNA–HDP (**b**) and pDNA–BAC (**c**).

In the case of the DNA–surfactant complex formation, the FT-MIR spectra show characteristic changes in the positions and/or intensities of some selected bands of molecular groups of both DNA and surfactant sensitive to moderate structural and environmental modifications. Taking into consideration the DNA and cationic surfactant molecules, the predicted complex formation is based on electrostatic interactions between nitrogen—containing a polar surfactant head group and the negatively charged phosphate group of the DNA backbone. As we expect, a shift in the band connected to the $\nu_{as}(PO_2^-)$ mode toward larger wavenumbers (above 1240 cm^{-1}) within the spectral range characteristic for A-type DNA conformation was observed in case of all three obtained complexes, which proves that this group is affected during complexation. Also, the band associated with the $\nu_{as}(PO_2^-)$ mode for the Z-DNA form and observed in pure pDNA at 1208 cm^{-1} shifted toward smaller wavenumbers by 7 cm^{-1} in case of pDNA–HDP and pDNA–BAC complexes. The bands connected to $\nu_s(CH_2)$ and $\nu_{as}(CH_2)$ stretching modes of hydrocarbon chains of all three surfactants (CTMA, HDP and BAC) shifted toward larger wavenumbers on the complex formation, indicating their slightly larger degree of disorder comparing to pure surfactants. This is consistent with our previous study [20]. The most significant changes in the IR band positions of the pDNA–surfactant complexes in comparison to pure components indicating molecular modifications occurring in the structure on complexation are summarized in Table 2.

Table 2. Comparison of the selected FT-MIR bands' positions (in cm^{-1}) for pure pDNA, CTMA, HDP, BAC and the pDNA–surfactant complexes with their tentative assignment based on the literature [23–25].

pDNA	CTMA	HDP	BAC	pDNA–CTMA	pDNA–HDP	pDNA–BAC	Assignment
	719 shp	717 shp	723 shp	724 br	722 br	725 br	$\rho(CH_2)_{al,n}$
801 ^{A}				800 ^{A}	800 ^{A}	801 ^{A}	Sugar—phosphate backbone
832 ^{B}				822 ^{B}	825 ^{B}	823 ^{B}	Sugar—phosphate backbone
860 ^{A}				–	–	–	Sugar—phosphate backbone
962 ^{B}				960	957	960	ν(C-C) of DNA backbone
~1015 sh				1011 shp	1009 sh	1010 shp	Furanose ring
1063				1062	1062	1062	ν(C-O) phosphodiester linkage
1088				1094	1092	1091	$\nu_s(PO_2^-)$
1176 ^{A}				1168 ^{A}	1172 ^{A}	1166 ^{A}	Sugar base
1208 ^{Z}				–	1201 ^{Z}	1201 ^{Z}	$\nu_{as}(PO_2^-)$
1238 ^{A}				1243 ^{A}	1246 ^{A}	1244 ^{A}	$\nu_{as}(PO_2^-)$
~1284 ^{B}				1274 ^{A}	1278 ^{A}	1278 ^{A}	Sugar base (T)
1418 ^{B}				1418 ^{B}	1415 $^{B\ or\ Z}$	1415 $^{B\ or\ Z}$	Sugar—phosphate backbone
	1471	1472 1508	1471 1490	1470 1485	1468 1504	1467 1484	$\nu(C{=}C)_{ar}$ $\nu(C{=}C)_{ar}$ or base ring
1531				1527	1528	1531	Deoxycytidine
1696				1690	1690	1689	Base double bond stretch
	2849	2849	2852	2854	2853	2854	$\nu_s(CH_2)_{al}$
	2917	2914	2921	2925	2925	2924	$\nu_{as}(CH_2)_{al}$

$^{A, B\ or\ Z}$ Positions of the IR bands characteristic for A, B or Z conformation of DNA. Mode description: ν—stretching; ρ—rocking; s—symmetric; as—asymmetric; ar—aromatic; al—aliphatic; n—long chain; sh—shoulder; shp—sharp; br—broad; T—thymine.

Summing up, the FT-MIR spectroscopic results are very consistent in the case of all the three obtained complexes, which indicates that the mechanism of their formation is the same and based on electrostatic interactions. The positions of the bands connected to the asymmetric phosphate (1243–1246 cm^{-1}) and phosphodiester (intense band at 1062 cm^{-1}) modes indicate that, upon complexation, most probably, the A-DNA conformation of double helix was predominant.

The diffraction patterns for the pDNA complexes (Figure 2) show a sharp peak at low angles $2\theta \approx 2.5$–2.7° and a wide maximum at about $2\theta \approx 20^\circ$. The latter corresponds to an average distance of 0.44–0.46 nm between the aliphatic chains of surfactants. The position of the peak at low angles corresponds to the distance 3.2–3.5 nm (Table 3), which can be interpreted as the distance $d_{02} = d_{11}$ in the two-dimensional centred rectangular lattice (see Figure 5 in [20]) with the unit cell parameters $a = b/\sqrt{3}$ and $b = 2d_{02}$. The condition $a = b/\sqrt{3}$, resulting in the overlapping of the (02) and (11) diffraction peaks, indicates the hexagonal packing of the DNA helices, with the distance between the nearest neighbours equal to a = 3.7–4.0 nm. The differences in the size of the unit cell arise from the different lengths of the surfactants' molecules.

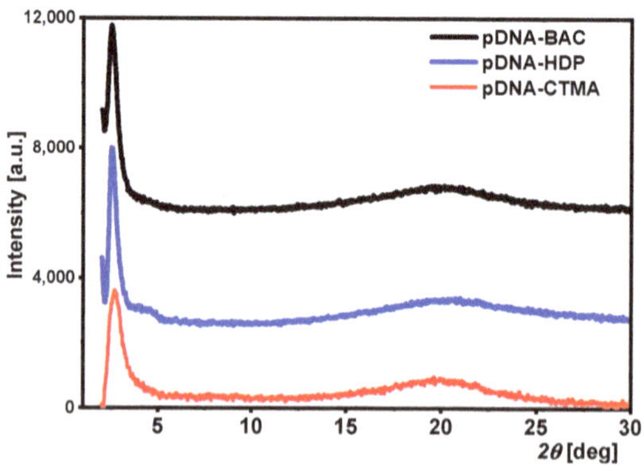

Figure 2. XRD patterns for pDNA-based complexes with three different surfactants.

Table 3. The maxima position for pDNA complexes and characteristic distances calculated on the basis of the Bragg equation.

	Low-Angle Peak		Peak at ca. 20°	
	2θ [deg]	d [nm]	2θ [deg]	d [nm]
pDNA–CTMA	2.77(1)	3.19(2)	19.69(3)	0.46(1)
pDNA–BAC	2.52(1)	3.50(2)	19.88(2)	0.45(2)
pDNA–HDP	2.60(1)	3.39(2)	20.54(2)	0.44(2)

The diffraction patterns recorded for the plasmid DNA complexes are similar to those obtained for the complexes synthesized based on linear DNA [20]. The structures formed by the circular and supercoiled forms of DNA are too large (the estimated hydrodynamic diameter of the circular plasmid form, approx. 4500 bp, is 320 nm [26]) to be observed in the diffraction patterns obtained by the standard XRD method, which allows the study of structures with sizes up to approx. 18 nm (distance corresponding to the angle $2\theta \approx 0.5^\circ$ for CuKα radiation).

As it was mentioned above, the dielectric properties of the pure surfactant and its complexes with linear DNA were described in our previous paper [20]. Here, we present and discuss the dielectric spectra of plasmid DNA complexes with three surfactants (Figure 3a,b). The smallest dielectric dispersion ε' and absorption ε'' as well as the smallest real part of specific electric conductivity σ' in the entire studied frequency range were recorded for the pDNA–BAC complex, while the highest belonged to the pDNA–HDP complex. Similarly, the smallest dielectric dispersion ε' was registered for the complex based on the linear DNA with the BAC surfactant (dsDNA–BAC) [20]. Thus, the type of surfactant, and/or its conformation, affects the dielectric dispersion ε' in complexes, similar to DNA solutions, where ε' depends on the type of cations and concentration (the higher the concentration of cations, the more phosphate groups are neutralized and the dielectric dispersion decreases) [6]. In turn, the dielectric absorption ε'' increases as the frequency decreases for all the pDNA-based complexes, and it is the highest for the pDNA–HDP complex and about one order of magnitude greater at the low frequency limit than for the pDNA–CTMA complex (Figure 3b). This behaviour is fundamentally different from that of the complexes based on linear DNA in which the dielectric absorption was constant over the entire frequency range and equal for both dsDNA–CTMA and dsDNA–HDP complexes [20]. Taking into account that the dielectric absorption for pDNA–HDP is about 3000 higher than that for dsDNA–HDP at the lowest frequency, it can be concluded that the type of DNA in the complex has a large impact on the dielectric properties.

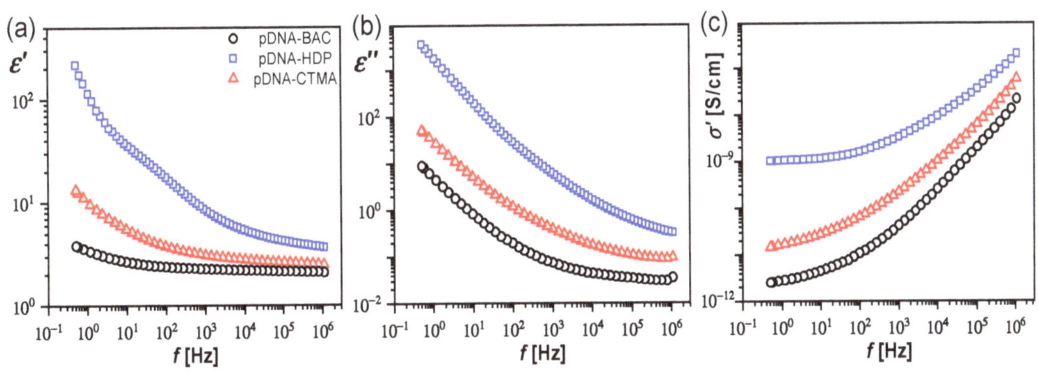

Figure 3. The dielectric dispersion (**a**) and absorption (**b**) as well as real part of specific electric conductivity (**c**) for pDNA complexes. The legend in (**a**) is the same for all graphs.

On the other hand, the real part of the specific electric conductivity σ' increases with increasing frequency for all the studied complexes, and, for selected frequencies (0.5 and 50.0 Hz), it is about three orders of magnitude lower for pDNA–BAC than for pDNA–HDP (Figure 3c, Table 4). Thus, the layer of pDNA–surfactant complexes can be used as a layer (or dielectric layer in capacitor) with a specific electric conductivity (or capacitance), which can be changed by selecting the appropriate surfactant. In addition, since the specific electric conductivity of pDNA complexes varies strongly with frequency (about three orders of magnitude), they can be used as a layer with a variable conductivity σ' by frequency selection (see Table 4).

Table 4. The dielectric dispersion and specific electric conductivity at two frequencies for the complexes studied.

	f = 0.5 Hz		f = 50.0 Hz	
	ε'	σ' [pS/cm]	ε'	σ' [pS/cm]
pDNA–BAC	3.85	2.55	2.44	7.68
pDNA–CTMA	13.25	14.57	4.11	47.79
pDNA–HDP	217.72	1019.8	21.87	1388.70

Since for the pDNA–HDP complex a weak relaxation process is visible around 10–200 Hz on the dielectric dispersion (Figure 3a), the dielectric spectra of all complexes studied ($\varepsilon'(f)$, $\varepsilon''(f)$) were transformed to the impedance representation ($Z'(f)$, $Z''(f)$) for its analysis. Figure 4 presents the real and imaginary parts of the impedance, while Nyquist plots are presented in Figure 5a–c. The relaxation process at the low frequency range is clearly visible for all the studied complexes.

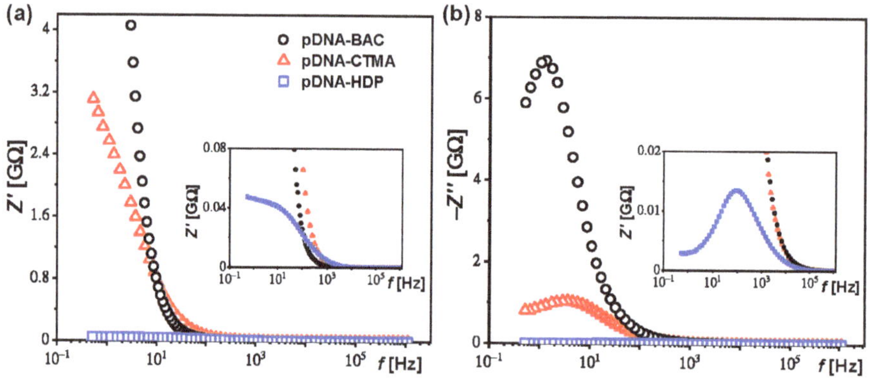

Figure 4. Real (**a**) and imaginary (**b**) part of the impedance for all complexes studied; legend in (**a**) is valid for (**b**).

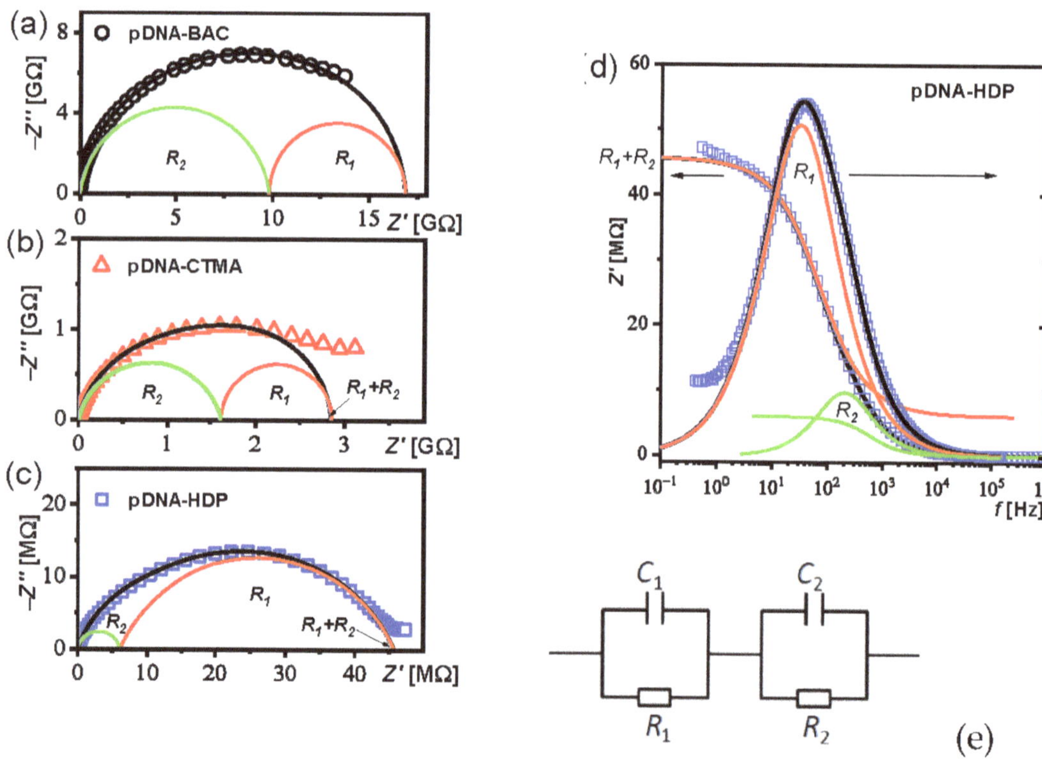

Figure 5. Nyquist plot for complexes studied: pDNA–BAC (**a**), pDNA–CTMA (**b**) and pDNA–HDP (**c**) as well as an impedance spectrum with fitting result for pDNA–HDP complex (**d**). Black line represents the best fit of Equation (1) to the experimental data (open points); the red line represents fitted data for the grain boundaries (R_1), while the green line represents the grain interior (crystallite part of the sample—R_2). Calculated equivalent circuit: two RC parallel connected in series (**e**).

The Cole–Cole model [27] was fitted to the experimental data. The measured complex impedance $Z^*(\omega)$ can be described by an equivalent circuit consisting of two parallel RC circuits connected in series:

$$Z^*(\omega) = \frac{R_1}{1 + (i\omega R_1 C_1)^{1-\alpha_1}} + \frac{R_2}{1 + (i\omega R_2 C_2)^{1-\alpha_2}} \quad (1)$$

where R_i is the sample direct current resistance of the *i*-th contribution; C_i is the electric capacity of the *i*-th contribution and α_i ($0 < \alpha \leq 1$) is a parameter of the relaxation time distribution, an empiric exponent of the *i*-th contribution related to the deviation from the Debye model. $\omega = 2\pi f$ is angular frequency, *f*—frequency of measuring AC electric field. The black line in Figure 5a–d represents the best fit of Equation (1) to the experimental data with the low- (red line—R_1) and high-frequency (green line—R_2) contributions. The fitting parameters obtained for all complexes studied are gathered in Table 5. As can be seen, the sum $R_1 + R_2$ was three orders of magnitude greater for pDNA–BAC (or pDNA–CTMA) than for the pDNA–HDP complex. A response consisting of two semicircles is often observed in polycrystalline samples, and it is related to two different contributions to total conductivity: conductivity of the crystallite part (high-frequency contribution—R_2) and the conductivity of the grain boundaries (low-frequency contribution—R_1) [28]. In our case, the samples are in the form of powder, so both contributions are possible. Based on the analysis presented in Figure 5, it can be concluded that the grain boundaries' contribution (R_1) exists for all the studied complexes based on plasmid DNA.

Table 5. Fitting parameters obtained from the impedance spectra analysis.

	pDNA–BAC	pDNA–CTMA	pDNA–HDP
R_1 [GΩ]	7.10 ± 0.07	1.24 ± 0.02	$(39.77 \pm 0.09) \cdot 10^{-3}$
τ_1 [s]	0.2204 ± 0.0049	0.0761 ± 0.0015	$(2.2321 \pm 0.0004) \cdot 10^{-3}$
C_1 [pF]	31.04 ± 0.25	61.35 ± 0.65	303.83 ± 1.18
α_1	0	0	0.277 ± 0.002
R_2 [GΩ]	9.81 ± 0.04	1.62 ± 0.01	$(5.93 \pm 0.05) \cdot 10^{-3}$
τ_2 [s]	0.0769 ± 0.0021	0.0181 ± 0.0005	$(2.6998 \pm 0.0002) \cdot 10^{-4}$
C_2 [pF]	9.65 ± 0.05	20.74 ± 0.15	124.86 ± 3.13
α_2	0.081 ± 0.001	0.154 ± 0.003	0.124 ± 0.004

As we have shown in our previous paper [20], the highest specific electric conductivity σ' in the whole frequency range occurred for the BAC surfactant, while, for CTMA and HDP, it was lower and comparable (which probably depends on the position of the nitrogen atom and/or the presence of an additional $-CH_2-$ group between the nitrogen atom and the benzene ring). The specific electric conductivity for the pDNA–CTMA and pDNA–HDP complexes increased relative to that of the pure surfactants, while for pDNA–BAC a slight decrease relative to pure BAC was visible. In turn, the σ' was the lowest for complexes with the BAC surfactant (linear dsDNA–BAC and plasmid pDNA–BAC), which may result from the conformation of the BAC chains when interacting with the negatively charged DNA double helix. This double helix has a negative charge coming from phosphoric acid (V) residues, where each phosphate residue has an unsaturated oxygen atom. Thus, negative charges on phosphate residues are the cause of electrostatic interaction (even the formation of weak covalent chemical bonds) between the DNA backbone and the cationic surfactant.

To find the origin of the second registered relaxation process (conductivity of the crystallite part, high-frequency contribution—R_2) in the pDNA–surfactant complexes studied, let us consider DNA solutions for a moment. In DNA solutions containing cations (counterions) such as Na^+ or Mg^{2+}, some of these cations are bound to the phosphate backbone by a weak chemical bond, while other cations are more loosely bound and can move around the helix. Cations attracted to the phosphate backbone form a cationic (counterionic) shell that shields part of the negative charge of the double helix. As a result, a double charge layer (the Stern layer) is formed [6]. The attracted cations show some mobility and oscillate around the phosphate charge centres in alternating electric fields (it is measuring field during experiments). These cationic oscillations represent a relaxation process that is visible at the low frequency range (1–100 Hz). In turn, the relaxation process associated with the movement of counterions bound to individual phosphate groups appears in the range of several MHz. Other relaxation processes occur at frequencies above hundreds of GHz, e.g., propeller twist relaxation, which occurs when two adjacent nitrogen bases in a pair twist in opposite directions, or breather relaxation, where two nitrogen bases oscillate in opposite to each other, which causes the hydrogen bonds to be alternately shortened and lengthened [6].

Although our study did not deal with DNA solutions but with pDNA–surfactant complexes in powder form, we believe that the charge double layer was formed in the same way and that the oscillating polar surfactant head groups were responsible for the relaxation process seen in at the higher frequency range. It is possible that during this relaxation process the aliphatic chain also oscillates/fluctuates, but we did not record any relaxation processes for pure surfactants [20]. The influence of the molecular weight of the surfactant on the relaxation time τ_2 of this process should also be considered (for the surfactant with the highest molecular weight (BAC), the relaxation time τ_2 should be the longest and the relaxation frequency the smallest, which we actually observed, see Table 5). Therefore, the shortest relaxation time should be for the pDNA–CTMA complex. However, the shortest relaxation time τ_2 was found for the pDNA–HDP complex. Thus, assuming that the relaxation process revealed at the high frequency range is of the same origin for all the studied pDNA-based complexes, the molecular weight of the cationic surfactant is not a dominant factor influencing the relaxation time of this process. It seems, however, that the conformation of the surfactant chain, which enforces the degree of packing, has a much greater impact. Finally, it cannot be ruled out that the relaxation time may also be affected by the purity of the surfactants and the residue of unwashed residual chloride ions (stabilizing pure surfactants).

4. Conclusions

Complexes based on plasmid DNA and three surfactants with aliphatic chains containing 16 carbon atoms (BAC, CTMA and HDP) in the powder form have been studied by Fourier-transform middle-infrared spectroscopy, X-ray diffraction and frequency-domain dielectric spectroscopy methods. Based on the results obtained, the following conclusions can be drawn:

- The DNA conformation in all three plasmid DNA complexes was found to be predominantly of the A type, although the plasmid DNA used to form the complexes was most likely a mixture of three conformations (A-DNA, B-DNA and Z-DNA) with B and A as the predominant forms. After complexation, the A-DNA double-helix form predominated in plasmid-based pDNA complexes, similarly to complexes based on linear dsDNA (although linear dsDNA with a predominant B-DNA form was used [20]).
- The pDNA helices in all studied complexes were locally packed into a hexagonal lattice, similarly to what was found in complexes of linear DNA with these surfactants (BAC, CTMA and HDP) [20]. The fact that the plasmid DNA was in a form of a closed loop did not prevent the hexagonal packing because the length of the used DNA (320 nm) was almost two orders of magnitude larger than the distances between neighbouring helices (3.7–4.0 nm).
- In contrast to linear DNA complexes studied by us earlier [20], two relaxation processes were revealed for plasmid DNA complexes (contributions to total conductivity coming from conductivity of the crystallites (grain interior) and from conductivity of the grain boundaries). The relaxation process in the crystallites revealed for all the studied complexes was interpreted as an oscillation of the polar surfactant head groups and seemed to be independent of the molecular weight of the cationic surfactant but dependent on the conformation of the surfactant chain.

- The dielectric absorption changed in a completely different way for complexes based on plasmid pDNA and linear dsDNA. It increased with decreasing frequency for all the pDNA-based complexes, whereas it was constant over the entire frequency range for all the dsDNA-based complexes [20]. Moreover, the dielectric absorption for pDNA–HDP was about 3000 higher than that for for dsDNA–HDP.
- The specific electric conductivity depended on the surfactant forming the complexes based on plasmid pDNA, as it was found earlier for linear dsDNA. In addition, it seems that the conformation of the BAC chains caused σ' to be the lowest for complexes with the BAC surfactant for both linear dsDNA and plasmid pDNA.

Summarizing, we showed that the type of the DNA (plasmid or linear) had a great influence on the dielectric properties of the complexes with cationic surfactants. The results of our study show that the specific electric conductivity of the layer of pDNA–surfactant complex can be changed by selecting the frequency, which is interesting from an application point of view. Considering that more and more attention is paid to recycling and biodegradability when designing new functional materials and that large-scale research into conductive biodegradable materials is currently underway, a complex based on plasmid DNA with a cationic surfactant seems to be a promising candidate. However, this issue requires further research, such as studying the influence of temperature and environment on the properties of these complexes.

Author Contributions: Conceptualization, M.M.; methodology, M.M.; validation, S.L., N.G., A.D. and M.M.; formal analysis, S.L., N.G., A.D. and J.Ś.; investigation, A.R., S.L., N.G. and A.D.; resources, M.M.; writing—original draft preparation, A.R., S.L., N.G., A.D. and M.M.; writing—review and editing, M.M.; visualization, S.L., A.D., N.G., J.Ś. and M.M.; supervision, M.M.; project administration, M.M.; funding acquisition, M.M. All authors have read and agreed to the published version of the manuscript.

Funding: This research was carried out with equipment purchased thanks to the financial support of the European Regional Development Fund in the framework of the Polish Innovation Economy Operational Program, Contract No. POIG.02.01.00–12–023/08 (Novocontrol spectrometer) and the European Regional Development Fund Operational Program Infrastructure and Environment, Contract No. POIS.13.01.00-00-062/08 (Empyrean 2 PANalytical diffractometer).

Institutional Review Board Statement: Not applicable.

Informed Consent Statement: Not applicable.

Data Availability Statement: Data are contained within the article and the Appendix A.

Acknowledgments: We would like to thank Teresa Jaworska-Gołąb (Institute of Physics, Jagiellonian University) for help and discussion regarding X-ray diffraction measurements and Robert Ekiert (Faculty of Biochemistry, Biophysics and Biotechnology, Jagiellonian University) for help and discussion regarding the preparation of research material. This research was supported in part by the Excellence Initiative—Research University Program at the Jagiellonian University in Kraków.

Conflicts of Interest: The authors declare no conflicts of interest.

Appendix A

FT-MIR Spectroscopy

Both FT-MIR spectra presented here were obtained at room temperature with a spectral resolution of 2 cm^{-1} and 32 scans per spectrum in transmission mode using a Bruker VERTEX 70v vacuum spectrometer. Bulk samples of the dried plasmid DNA and linear DNA were mixed with KBr, compressed into pellets and measured. Additionally, parallel to these measurements, the FT-ATR reference spectra were collected for comparison to ensure that the compression did not damage the DNA material.

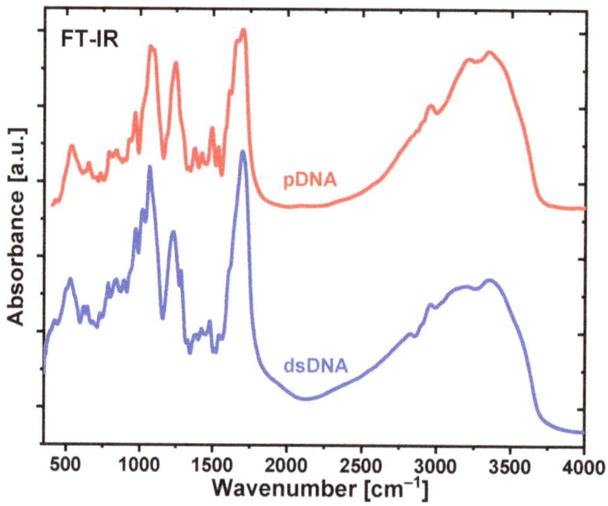

Figure A1. Comparison of FT-MIR spectra of bulk plasmid DNA (pDNA) and linear DNA (dsDNA). Results for linear DNA are presented in [20].

References

1. Watson, J.D.; Crick, F.H.C. Molecular Structure of Nucleic Acids: A Structure for Deoxyribose Nucleic Acid. *Nature* **1953**, *171*, 737. [CrossRef] [PubMed]
2. Cooper, G.M.; Hausman, R.E. *The Cell: A Molecular Approach*, 6th ed.; Sinauer Associates: Sunderland, MA, USA, 2007; Volume 4.
3. Ghosh, A.; Bansal, M. A glossary of DNA structures from A to Z. *Acta Crystallogr. Sect. D Biol. Crystallogr.* **2003**, *59*, 620. [CrossRef] [PubMed]
4. Kado, C. Origin and evolution of plasmids. *Antonie Leeuwenhoek* **1998**, *73*, 117. [CrossRef] [PubMed]
5. Mielke, C.; Benham, S. DNA mechanics. *Annu. Rev. Biomed. Eng.* **2005**, *7*, 21.
6. Baker-Jarvis, J.; Jones, C.A.; Riddle, B. *Electrical Properties and Dielectric Relaxation of DNA in Solution*; NIST Technical Note 1509; NIST: Boulder, CO, USA, 1998.
7. Heckman, E.M.; Hagen, J.A.; Yaney, P.P.; Grote, J.G.; Hopkins, F.K. Processing techniques for deoxyribonucleic acid: Biopolymer for photonics applications. *Appl. Phys. Lett.* **2005**, *87*, 211115. [CrossRef]
8. Steckl, A.J. DNA—A new material for photonics? *Nat. Photonics* **2007**, *1*, 3. [CrossRef]
9. Yumusak, C.; Singh, T.B.; Sariciftci, N.S.; Grote, J.G. Bio-organic field effect transistors based on crosslinked deoxyribonucleic acid (DNA) gate dielectric. *Appl. Phys. Lett.* **2009**, *95*, 263304. [CrossRef]
10. Kwon, Y.-W.; Lee, C.H.; Choi, D.-H.; Jin, J.-I. Materials science of DNA. *J. Mater. Chem.* **2009**, *19*, 1353. [CrossRef]
11. Singh, T.B.; Sariciftci, N.S.; Grote, J.G. Bio-Organic Optoelectronic Devices Using DNA. In *Organic Electronics*; Advances in Polymer Science; Grasser, T., Meller, G., Li, L., Eds.; Springer: Berlin/Heidelberg, Germany, 2010; Volume 223, p. 73.
12. Chi, Q.; Yang, Z.; Xu, K.; Wang, C.; Liang, H. DNA Nanostructure as an Efficient Drug Delivery Platform for Immunotherapy. *Front. Pharmacol.* **2020**, *10*, 1585. [CrossRef]
13. Mirkin, C.A.; Letsinger, R.L.; Mucic, R.C.; Storhoff, J.J. A DNA-based method for rationally assembling nanoparticles into macroscopic materials. *Nature* **1996**, *382*, 607. [CrossRef]
14. Braun, E.; Eichen, Y.; Sivan, U.; Ben-Yoseph, G. DNA-templated assembly and electrode attachment of a conducting silver wire. *Nature* **1998**, *391*, 775. [CrossRef] [PubMed]
15. Kawabe, Y.; Wang, L.; Horinouchi, S.; Ogata, N. Amplified Spontaneous Emission from Fluorescent-Dye-Doped DNA–Surfactant Complex Films. *Adv. Mater.* **2000**, *12*, 1281. [CrossRef]
16. Singh, B.; Sariciftci, N.S.; Grote, J.G.; Hopkins, F.K. Bio-organic-semiconductor-field-effect-transistor based on deoxyribonucleic acid gate dielectric. *J. Appl. Phys.* **2006**, *100*, 024514. [CrossRef]
17. Yukimoto, T.; Uemura, S.; Kamata, T.; Nakamura, K.; Kobayashi, N. Non-volatile transistor memory fabricated using DNA and eliminating influence of mobile ions on electric properties. *J. Mater. Chem.* **2011**, *21*, 15575. [CrossRef]
18. LaBoda, C.; Duschl, H.; Dwyer, C.L. DNA-Enabled Integrated Molecular Systems for Computation and Sensing. *Acc. Chem. Res.* **2014**, *47*, 1816–1824. [CrossRef] [PubMed]
19. Benenson, Y.; Adar, R.; Paz-Elizur, T.; Livneh, Z.; Shapiro, E. DNA molecule provides a computing machine with both data and fuel. *Proc. Natl. Acad. Sci. USA* **2003**, *100*, 2191–2196. [CrossRef] [PubMed]
20. Radko, A.; Lalik, S.; Deptuch, A.; Jaworska-Gołąb, T.; Ekiert, R.; Górska, N.; Makyła-Juzak, K.; Nizoł, J.; Marzec, M. Physicochemical characterization of the DNA complexes with different surfactants. *Polymer* **2021**, *235*, 124277. [CrossRef]

21. Radko, A.; Nizioł, J.; Makyła-Juzak, K.; Ekiert, R.; Górska, N.; Górecki, A.; Marzec, M. Properties of DNA-CTMA monolayers obtained by Langmuir-Blodgett technique. *Mater. Sci. Eng. B* **2021**, *263*, 114859. [CrossRef]
22. Roisnel, T.; Rodriguez-Carvajal, J. WinPLOTR: A Windows tool for powder diffraction patterns analysis. In Proceedings of the Seventh European Powder Diffraction Conference, Barcelona, Spain, 20–23 May 2000.
23. Taillandier, E.; Peticolas, W.L.; Adam, S.; Huynh-Dinh, T.; Igolen, J. Polymorphism of the d(CCCGCGGG)2 double helix studied by FT-IR spectroscopy. *Spectrochim. Acta* **1990**, *46A*, 107–112. [CrossRef]
24. Taillandier, E.; Liquier, J. Infrared spectroscopy of DNA. *Methods Enzymol.* **1992**, *211*, 307–335.
25. Sofińska, K.; Wilkosz, N.; Szymoński, M.; Lipiec, E. Molecular spectroscopic markers of DNA damage. *Molecules* **2020**, *25*, 561. [CrossRef] [PubMed]
26. Arkhangelskya, E.; Sefib, Y.; Hajajb, B.; Rothenbergc, G.; Gitis, V. Kinetics and mechanism of plasmid DNA penetration through nanopores. *J. Membr. Sci.* **2011**, *371*, 45–51. [CrossRef]
27. Cole, K.S.; Cole, R.H. Dispersion and Absorption in Dielectrics I. *J. Chem. Phys.* **1941**, *9*, 341. [CrossRef]
28. Michalska, M.; Ławniczak, P.; Strachowski, T.; Ostrowski, A.; Bednarski, W. Structural studies and selected physical investigations of $LiCoO_2$ obtained by combustion synthesis. *Beilstein J. Nanotechnol.* **2022**, *13*, 1473–1482. [CrossRef]

materials

Article

Dielectric Modes in Antiferroelectric and Ferroelectric Liquid Crystals in a Pure Enantiomeric Version and a Racemic Mixture

Paweł Perkowski [1,*] and Magdalena Urbańska [2]

[1] Institute of Applied Physics, Military University of Technology, Kaliskiego 2, 00-908 Warsaw, Poland
[2] Institute of Chemistry, Military University of Technology, Kaliskiego 2, 00-908 Warsaw, Poland; magdalena.urbanska@wat.edu.pl
* Correspondence: pawel.perkowski@wat.edu.pl

Abstract: The dielectric properties of synclinic (ferroelectric SmC*) and anticlinic (antiferroelectric SmC_A*) smectic liquid crystals composed of molecules of one chiral version (S) are presented and compared with properties of racemic mixture (R, S), showing SmC and SmC_A phases. The racemic mixture completely loses its ferroelectric and antiferroelectric properties. Surprisingly, only one dielectric mode observed in the antiferroelectric SmC_A* phase disappeared in the dielectric response of the racemic SmC_A phase. Additionally, we observed that in the SmC phase, seen in the racemic mixture, the weak dielectric mode (named the X mode) is detected, which seems to be the continuation of the P_L mode existing in the racemic SmC_A. Moreover, this mode in the racemic SmC has nothing to do with the Goldstone mode, typical for the SmC* phase. This paper describes in detail the real and imaginary parts of dielectric permittivity in smectic phases for the enantiomer and racemate with and without a DC field, compares the properties of the X and P_L modes, and discusses the full scheme of dielectric modes in enantiomer and racemate.

Keywords: antiferroelectric liquid crystal; ferroelectric liquid crystal; (S) enantiomer; racemate; chirality; dielectric spectroscopy; relaxation mode; activation energy; Arrhenius law

1. Introduction

The definition of chirality was coined by William Thomson (Lord Kelvin) in his "Robert Boyle Lecture" at Oxford University in 1893. He wrote the following in his "*Baltimore Lectures on Molecular Dynamics and the Wave Theory of Light*" in 1904 [1]: "I call any geometrical figure or group of points chiral and say it has chirality, if its image in a plane mirror, ideally realized, cannot be brought to coincide with itself. Two equal and similar right hands are homochirally similar. Equal and similar right and left hands are heterochirally similar".

Chirality is an important factor in life. Indeed, all amino acids (except glycine) that build the proteins in our bodies are chiral molecules. Moreover, natural amino acids appear only in one chiral form. Our nose can recognize the chirality of heterochirally similar natural substances. Additionally, medicine in one chiral form can cure while the opposite form of the same medicine can be poisonous.

In general, chirality plays an important role in organic chemistry. Chirality is the main feature of molecules building liquid crystals, which leads to the polar properties of ferro-, antiferro-, and ferrielectricity in mesophases and the creation of helical superstructures. The source of polar properties differs from "proper" polar phases. In the "proper" polar phases, spontaneous polarization results from the effect of a high density of strong dipole moments that interact electrically with each other [2–4], as is observed in nematic ferroelectric liquid crystals. In the "improper" polar phases, spontaneous polarization in tilted smectic phases results from chirality and mirror symmetry breaking [5,6]. Chirality introduced into the molecular structure makes the rotational potential around the long molecular axis asymmetric [7–10]. Hence, spontaneous polarization appears in the direction perpendicular to the tilt plane in the

SmC* and SmC_A* phases. After the discovery of liquid crystals by Reinitzer in 1888 [11], it took 90 years to discover improper ferroelectricity in 1978 in DOBAMC [12]. It took another 10 years until improper antiferroelectricity was found in liquid crystals [13]. Additionally, another 30 years passed before proper ferroelectricity was found in nematics in 2016 [2–4,14].

The dielectric response of the polar phase strongly differs from the dielectric response of the racemic version of the investigated liquid crystal. The best example is the strong Goldstone mode visible in the ferroelectric SmC* phase [15,16]. In the racemic analog of the SmC* phase (racemic SmC), this strong mode is not detectable [17–19] because of the disappearance of the helical superstructure.

The antiferroelectric SmC_A* phase exhibits two collective modes, P_H and P_L, at room temperature, as described in many papers [20–26]. These are called non-cancelled modes because they exist only for non-perfect anticlinic double-layer structures. The director in the first layer is not exactly on the opposite side of the cone, in comparison with the director in the next smectic layer, because of the small rotation caused by the helical superstructure. Hence, net polarization can exist in the double-layer unit. In a perfect double-layer structure, both modes would be canceled and non-active in dielectric response [15]. The P_H mode is faster and is called the anti-phase phason mode. For this collective mode, molecules in neighboring layers rotate (on the cone) in opposite directions. The P_L mode is slower and is called the in-phase phason mode. For this collective mode, molecules in neighboring layers rotate (on the cone) in the same direction. Additionally, one molecular mode called the S mode (molecular rotation around the short molecular axis), faster than the P_H mode, is also visible at room temperature, while the molecular L mode (molecular rotation around the long molecular axis) is too fast to be detectable at room temperature using standard impedance analyzers. When the antiferroelectric phase nucleates from the ferroelectric phase at cooling, one can notice that the amplitude of the Goldstone mode in the SmC* phase is dozens of times stronger than those of modes P_H and P_L [27]. As mentioned above, this results from the fact that the polarizations from adjacent layers in the SmC_A* phase are close to being canceled, while in the SmC* phase, they add up.

The DC field is often used to investigate electrically active modes. It can modify the dielectric response of ferroelectric and antiferroelectric phases. The DC field suppresses the Goldstone mode in the SmC* phase because it blocks the molecular movement on the cone by the unwinding of the helix [28]. In the SmC_A* phase, the DC influence is a little different. Two modes (P_H and P_L) are gradually strengthened by increasing the low DC field, while the S mode is weakened. When the DC field is high enough, both collective modes disappear because of the cancellation of the helical superstructure. For the suppression of both the P_H and P_L modes, we need a higher DC field than for the suppression of the Goldstone mode. Hence, when we suppress (using the DC field) the Goldstone mode at the SmC*-SmC_A* phase transition, we can see both antiferroelectric modes, which are normally hidden by the ferroelectric mode [29,30].

Another way to suppress the Goldstone mode as a sign of ferroelectricity is via racemization. When we use this method, we can see the dielectric response usually hidden by strong Goldstone mode. Earlier investigations [17–19] show that the dielectric response of a racemic mixture is richer than we can expect, and some modes in racemate behave as modes in an enantiomer.

This paper's main goal is to answer the question of how racemization influences the dielectric response of anticlinic SmC_A and synclinic SmC phases in comparison with the dielectric response of enantiomer and its anticlinic SmC_A* and synclinic SmC* phases.

The symbols used in this paper include the following:

SmC* —ferroelectric synclinic phase with a helical superstructure and its axis of rotation perpendicular to the smectic layers. Molecules in one smectic layer are tilted by the angle θ.

SmC_A* —antiferroelectric synclinic phase with a helical superstructure and its axis of rotation perpendicular to the smectic layers. Molecules in one smectic layer are tilted by the angle θ, while molecules in the next layer are tilted by the angle $-\theta$.

SmC —synclinic phase without a helical superstructure. Molecules in one smectic layer are tilted by the angle θ. The helical superstructure and ferroelectricity vanish when the SmC phase is built from achiral molecules or SmC is the racemate—the phase is a mixture of 50% S-enantiomer and 50% R-enantiomer.

SmC_A —anticlinic phase without a helical superstructure. Molecules in one smectic layer are tilted by the angle θ, while molecules in the next layer are tilted by the angle $-\theta$. The helical superstructure and antiferroelectricity vanish when the SmC_A phase is built from achiral molecules or SmC_A is the racemate—the phase is an equimolar mixture of S- and R-enantiomers.

Goldstone mode —collective and strong dielectric mode (molecules move collectively around the cone in the smectic layer) in the same direction. During the rotation the phase angle changes in time, so this mode is called the phason mode.

S mode —molecular dielectric mode when the molecule rotates around its short molecular axis. This mode is observed in isotropic liquid, nematic, or smectic phases. This mode is Arrhenius type: its relaxation frequency decreases with decreasing temperature. The relaxation frequency of the S mode is 1k–10k times lower than the relaxation frequency of the L mode.

L mode —molecular dielectric mode when the molecule rotates around its long molecular axis. This mode is observed in isotropic liquid, nematic, or smectic phases. This mode is also Arrhenius type. It is not usually observed in standard impedance spectroscopy because it is fast.

P_H mode —collective and weak (in comparison with the Goldstone mode) dielectric mode detectable in antiferroelectric anticlinic SmC_A^*. It is also Arrhenius type. Molecules in neighboring layers, in the two-layer unit, rotate in opposite directions around the cone; hence, it is called the anti-phase phason mode. The relaxation frequency of the P_H mode is 1k times lower than the relaxation frequency of the S mode and 1k times higher than the relaxation frequency of the P_L mode.

P_L mode —collective and weak (in comparison with Goldstone mode) dielectric mode detectable in antiferroelectric anticlinic SmC_A^*. It is also Arrhenius type. Molecules in neighboring layers, in the two-layer unit, rotate in the same direction around the cone; hence, it is called the in-phase phason mode. The relaxation frequency of the P_L mode is 1k lower than the relaxation frequency of the P_H mode.

X mode —the dielectric mode described in this paper, which is observed in the racemic synclinic SmC phase. It seems to be the continuation of the P_L mode from the racemic SmC_A phase. Its strength at the SmC-SmC_A phase transition is twice weaker than the strength of the P_L mode close to the SmC_A-SmC phase transition.

$\delta\varepsilon_i$ —the strength of the *i*th dielectric mode.

f_{ri} —the relaxation frequency of the *i*th dielectric mode.

ε_∞ —high-frequency limit of permittivity.

ε_S —low-frequency limit of permittivity.

α_i and β_i —Havriliak–Negami distribution parameters.

j —imaginary unit.

σ_{ion} —ionic conductivity of the sample.

n —parameter describing ion contribution to the imaginary part of permittivity (close to one).

C —capacity in the parallel equivalent circuit.

G —conductivity in the parallel equivalent circuit.

C_0 —empty cell capacity.

f_∞ —limit of relaxation frequency at high temperatures (in Arrhenius law).

k —Boltzmann constant.

T —temperature on the thermodynamic scale.

2. Materials and Methods

2.1. Studied Compound

To show how racemization influences the dielectric response of the antiferroelectric SmC_A* and ferroelectric SmC* phases, the monofluorinated compound [31] and its racemic mixture [32] (Figure 1) were studied using dielectric spectroscopy. Importantly, the pure chiral compound (S) and its racemic mixture (R, S) exhibit both the anticlinic and synclinic phases.

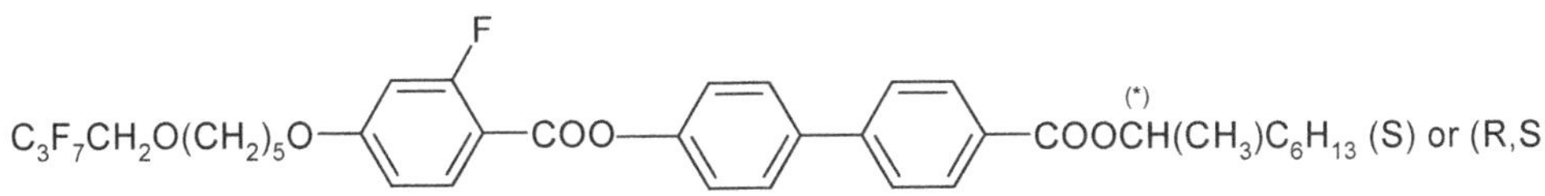

Figure 1. The molecular structure of the investigated liquid crystal. The asterisk (*) indicates where the chiral carbon atom (center) in the molecular structure is.

2.2. DSC Measurements

The transition temperatures and enthalpies of the transition for the enantiomer and racemate were determined by differential scanning calorimetry (DSC) using a SETARAM 141 microcalorimeter (KEP Technologies, Caluire-Et-Cuire, France) at the heating/cooling rate of 2 °C/min. The weight of each sample was about 20–30 mg.

2.3. Microscopy Observations

Mesophases were identified by observing the textures using the OLYMPUS BX51 optical microscope (Olympus Co., Tokyo, Japan) under crossed polarizers with the Linkam THMS-600 hot stage (Linkam Scientific Instruments LTD, Salfords, UK) controlled by the Linkam TMS-93 temperature programmer at the heating/cooling rate of 2 or 3 °C/min.

2.4. Dielectric Spectroscopy

Impedance spectroscopy is a useful method to characterize the dielectric properties of liquid crystals. The HP 4192A impedance analyzer (Hewlett Packard Inc., Palo Alto, CA, USA) was used in our investigations. This equipment allows, nominally, to take measurements at frequencies from 5 Hz to 13 MHz. Our previous experience with liquid crystal research led us to conclude that the measurement range (with acceptable accuracy) is narrower from 100 Hz to 10 MHz. At lower frequencies, ions distort the measurements, while at higher frequencies, the measurements are influenced by well-known high-frequency parasitic effects [33]. Our impedance analyzer enables applying the DC field from 0 V to 30 V, and it does not influence the measuring procedure at different AC signals. Hence, the DC field influence on the dielectric properties of the measured medium can be easily observed. In our investigation, the sign of the DC field was not important.

For measurements, self-made cells with gold electrodes (with an area of 25 mm^2), to minimize high-frequency parasitic effects, were used. Such cells can be used for frequencies up to 10 MHz. The thickness of the used cells was around 5 μm, while the cell's alignment was planar. The aligning layer was polyimide SE130 (Nissan Chemical Corporation, Tokyo, Japan). Liquid crystals were heated and put in a measuring cell in the isotropic phase using capillary action. Six measuring cycles were performed to understand the dielectric properties fully. The measurements were performed without a DC field and under 5 V or 10 V DC fields for the racemic mixture and pure enantiomer. In this paper, the results obtained at the 5 V DC field are not presented. All measurements were performed on the cooling cycles (starting from isotropic liquid). To control the temperature, the Linkam THMS-600 hot stage (Linkam Scientific Instruments LTD, Salfords, UK) controlled by a Linkam TMS-92 was used. The cooling rate was 0.5 °C/min.

The measurements were performed in a parallel equivalent circuit as follows: capacity C together with the conductivity G. Knowing the circuit parameters versus frequency $C(f)$ and $G(f)$, the real ε' and imaginary ε'' parts of the permittivity of the investigated compounds can be calculated using simple formulas as follows: $\varepsilon'(f) = C(f)/C_0$ and $\varepsilon''(f) = G(f)/2\pi f C_0$, where C_0 stands for empty cell capacity. C_0 was measured in an experimental setup just before filing the cell with liquid crystal.

Usually, the pure results of the dielectric spectroscopy of liquid crystals obtained in experiments are presented in the following four ways: (1) the real part ε' of the permittivity plot (at several frequencies) vs. temperature, (2) the imaginary part ε'' of the permittivity plot (at several frequencies) vs. temperature, (3) the real part ε' of the permittivity plot (at several temperatures) vs. frequency (frequency spectrum), and (4) the imaginary part ε'' of the permittivity plot (at several temperatures) vs. frequency (frequency spectrum). The first and second ways are preferred when the confirmation of existing liquid crystal phases is important. The third and fourth ways are suitable when the dielectric properties of each phase are analyzed. In this paper, we use the first and third methods of presentation.

2.5. The Method for Calculating the Parameters of Dielectric Modes

The main problem related to the dielectric spectroscopy of liquid crystals is knowing the true permittivity of liquid crystals. Parasitic effects make the permittivity measured in the experiment not exactly equal to the permittivity of liquid crystal [34]. This means that true values of liquid crystal permittivity should be calculated from experimental values. Fortunately, when cells with gold electrodes are used in measurements. the difference between the permittivity measured in the experiment and the true permittivity of a liquid crystal is small. The whole numerical procedure used in our calculations is explained in [35]. Important parameters in this procedure are the cut-off frequency of the *RC* measuring circuit and the resonance frequency of the *LC* measuring circuit.

Knowing the true permittivity of a given liquid crystal, one can calculate the parameters of the dielectric modes observed in an experiment. For the calculation of complex liquid crystal permittivity, Formula (1) was used (the Havriliak–Negami model [36]).

$$\varepsilon^* = \varepsilon' - j\varepsilon'' = \varepsilon_\infty + \sum_{i=1}^{3} \frac{\delta\varepsilon_i}{\left[1 + (jf/f_{ri})^{1-\alpha_i}\right]^{\beta_i}} - j\frac{\sigma_{ion}}{(2\pi f)^n C_0}, \tag{1}$$

where ε_∞ stands for the high-frequency limit of permittivity, $\delta\varepsilon_i$ stands for the strength of *i*th mode, f_{ri} stands for relaxation frequency of *i*th mode, α_i and β_i stand for the Havriliak–Negami distribution parameters, *j* stands for an imaginary unit, σ_{ion} stands for the ionic conductivity of the sample, *n* stands for a parameter (close to one), and C_0 stands for empty cell capacity. Calculations were performed for different phases. The maximum number *i* of the considered modes was 3.

The most important parameters found using Equation (1) are the relaxation frequencies f_{ri} and strengths $\delta\varepsilon_i$ of the analyzed modes. Knowing the relaxation frequencies of analyzed modes vs. temperature, the activation energy E_a from the Arrhenius law (2) can be calculated [37]. Of course, the relaxation frequency of the analyzed mode should follow the Arrhenius law.

$$f_r(T) = f_\infty \exp\left(-\frac{E_a}{kT}\right), \tag{2}$$

where f_r stands for relaxation frequency, f_∞ stands for the limit of relaxation frequency at high temperatures, *k* stands for Boltzmann constant, and *T* stands for the temperature on the thermodynamic scale.

3. Results

3.1. Results of the DSC Measurements and Microscopic Observations Performed for the Pure Enantiomer and Racemate

From DSC and polarizing optical microscopy, one can find that the pure enantiomer (in both the S and R forms) exhibits the following phase transition scheme: Cr 28.1 °C SmC_A* 99.0 °C SmC* 100.2 °C SmA* 101.1 °C Iso. Meanwhile, the racemic mixture manifests a slightly different phase sequence: Cr 39.1 °C SmC$_A$ 84.3 °C SmC 95.1 °C Iso. The DSC plots for racemate and enantiomer are presented in the Supplementary Materials.

The textures of liquid crystalline phases for the enantiomer and the racemate are also different, as shown in Figures 2 and 3. The width of all the microphotographs is about 600 μm. In the enantiomeric version, textures in both the SmC* and SmC$_A$* phases manifest clear dechiralization lines (Figure 2b,c) because of the existence of a helical superstructure. When the racemate is observed, the SmC phase shows no dechiralization lines (Figure 3a). Surprisingly, in the SmC$_A$ phase (Figure 3b), one can see weak and fuzzy dechiralization lines. This suggests that the compensation in the prepared racemate is not full.

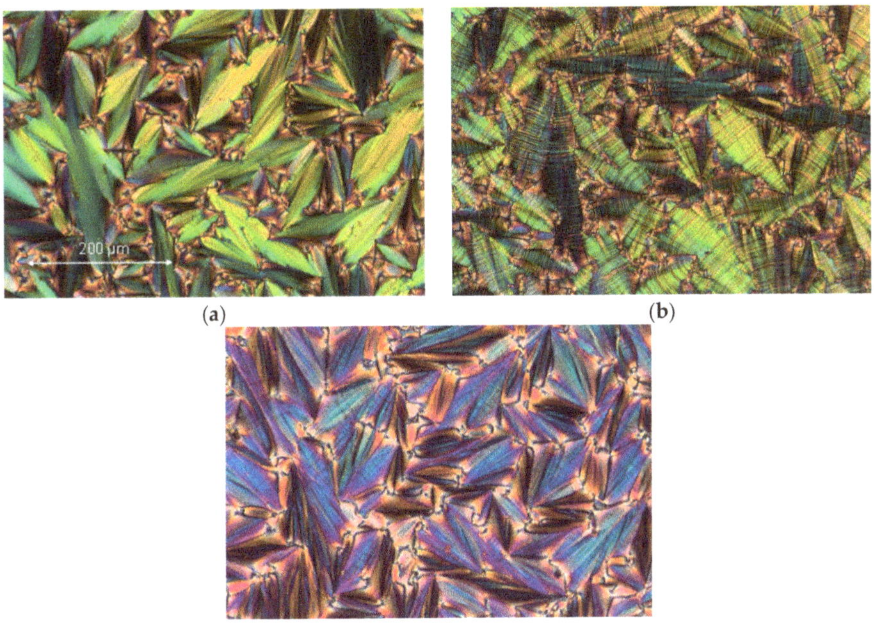

(**a**) (**b**) (**c**)

Figure 2. Photos showing the microscopic pattern of the racemate taken during the cooling cycle: (**a**) the SmA* phase at 98.5 °C, (**b**) the SmC* phase at 98.1 °C, and (**c**) the SmC$_A$* phase at 88.0 °C.

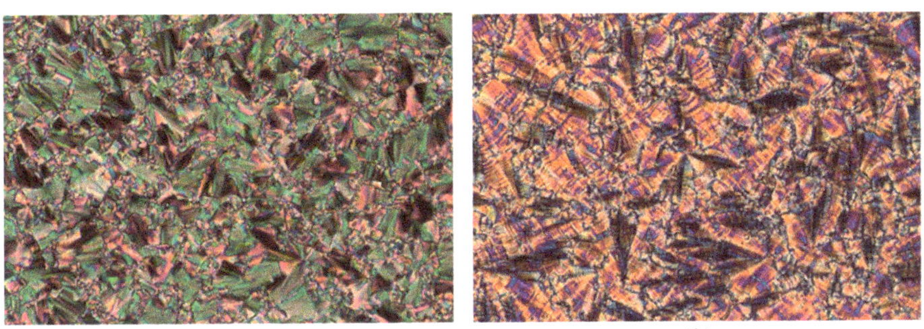

(**a**) (**b**)

Figure 3. Photos showing the microscopic pattern of the racemate taken during the cooling cycle: (**a**) the SmC phase at 94.7 °C and (**b**) the SmC$_A$ phase at 76.8 °C.

3.2. Results of the Measurements Performed for the Pure Enantiomer

Figure 4 shows the real part $\varepsilon'_{\perp}$ of permittivity (measured in a planarly aligned cell) versus temperature for twelve frequencies. This way of presenting dielectric properties is convenient for showing how the dielectric response changes when different phases appear with temperature. It is obvious when the dielectric properties exhibit dispersion. One can see poor dispersion in isotropic liquid and a strong dielectric mode, known as the Goldstone mode (G), seen in the SmC* phase (ferroelectric phason mode). The maximum permittivity value in the SmC* is around 170 (at 100 °C). The mesophase below the SmC* exhibits lower permittivity (the permittivity is comparable to isotropic liquid's permittivity: ~6). Three modes are seen in this phase. They are the P_L, P_H, and S modes in the SmC_A* phase. The first mode, P_L, is called the in-phase antiferroelectric phason mode, the second mode, P_H, is called the anti-phase antiferroelectric phason mode, and the last S mode is called the molecular mode around the short molecular axis. Both phases, SmC* and SmC_A*, seem to behave classically, as described in the literature [18,20,21,23,24].

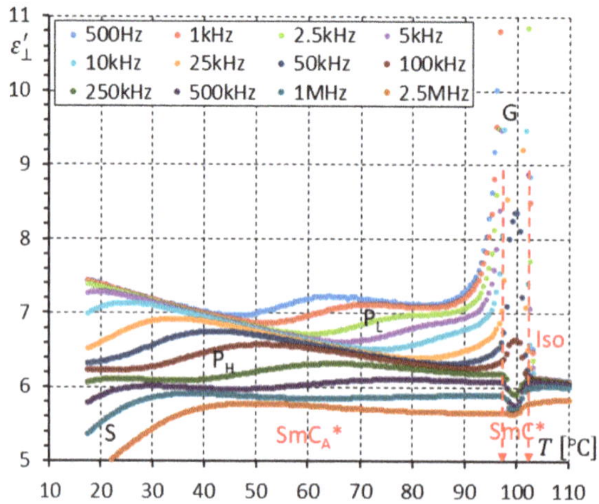

Figure 4. The real part $\varepsilon'_{\perp}$ of permittivity for twelve measuring frequencies versus temperature T for the enantiomer (no DC field applied). The vertical scale allows us to see the permittivity in the SmC* phase (G—the Goldstone mode) and the permittivity in the SmC_A* phase (three modes (P_L, P_H, S)) simultaneously. Arrows indicate phase transitions.

Figures 5 and 6 show the imaginary part $\varepsilon''_{\perp}$ of permittivity versus frequency for several temperatures in both the SmC* and SmC_A* phases (Figure 5) and in SmC_A* only (Figure 6). In Figure 5, the Goldstone mode (G) gradually disappears with decreasing temperature, while both the P_L and P_H modes gradually appear. Alternatively, it can be said that the weak P_L and P_H modes that exist in the dielectric response of the SmC* phase (close to the SmC*-SmC_A* phase transition) become gradually visible. We see that the Goldstone mode and the P_L and P_H modes coexist in the 97–91 °C temperature range. It is difficult to indicate precisely the transition temperature SmC*-SmC_A*. This phase transition (second rank) is a continuous process.

In Figure 6, only the spectra for the antiferroelectric (SmC_A*) phase are shown. Three modes are seen with their relaxation frequencies well separated in the frequency domain. When the temperature decreases, all relaxation frequencies also decrease (see arrows in Figure 6). At room temperature (25 °C), the S mode relaxation frequency is around 3.5 MHz, the P_H mode is simultaneously around 28 kHz, while the P_L mode is below 100 Hz. All three modes are Arrhenius-like. The P_H mode becomes stronger when the temperature decreases. The same effect should be also seen for the P_L and S modes. It is not visible because of the ion contribution (for the P_L mode) and high-frequency parasitic effects (for the S mode) [33]. Owing to the ion contribution, any existing modes at frequencies below the relaxation frequency of the P_L mode can not be seen.

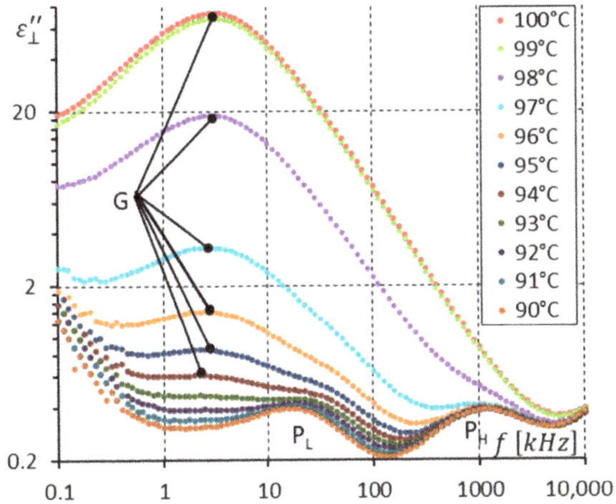

Figure 5. The imaginary part $\varepsilon''_{\perp}$ of permittivity for eleven temperatures (in the SmC* and SmC_A* phases, close to the SmC*-SmC_A* phase transition) versus measuring frequency f for the enantiomer (no DC field applied).

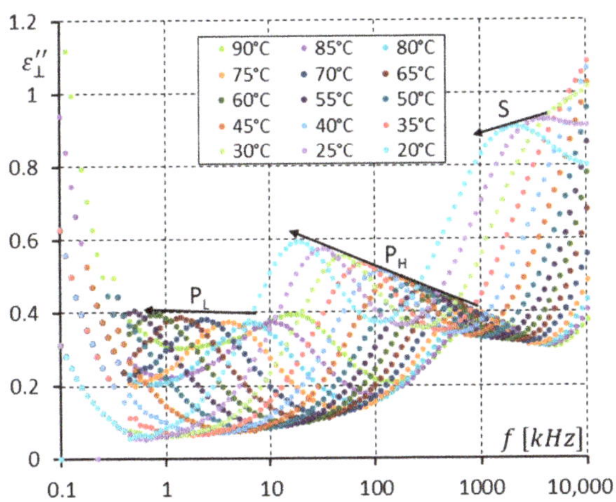

Figure 6. The imaginary part $\varepsilon''_{\perp}$ of permittivity for fifteen temperatures (in the SmC_A* phase only) versus measuring frequency f for the enantiomer (no DC field applied). Three relaxations are seen. Arrows show how the maximum amplitude of $\varepsilon''_{\perp}$ (also the relaxation frequencies) shifts when the temperature decreases.

After applying the DC field (10 V) during measurement (Figures 7–9), the dielectric response changes in the way one would predict. The Goldstone mode in the SmC* phase is almost suppressed, while all modes in the SmC_A* phase change slightly. In Figure 7, the real part of permittivity $\varepsilon'_{\perp}$ for several measuring frequencies is shown versus temperature. One can see that the strengths of the P_L and P_H modes increase under the DC field (compare Figures 4 and 7).

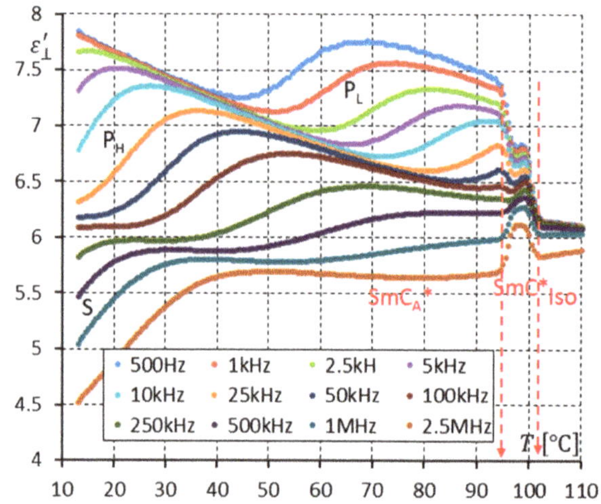

Figure 7. The real part $\varepsilon'_\perp$ of permittivity for twelve measuring frequencies versus temperature T for the enantiomer (10 V DC field applied). Three modes (P_L, P_H, S) are seen in the SmC$_A$* phase, while the Goldstone mode is fully suppressed in the SmC* phase. Arrows indicate the phase transitions.

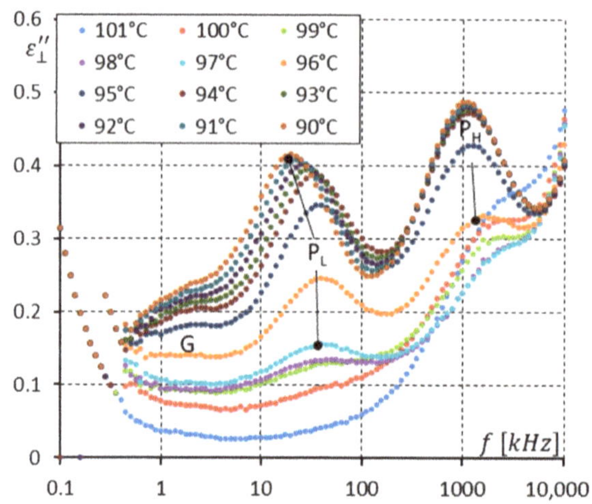

Figure 8. The imaginary part $\varepsilon''_\perp$ of permittivity for twelve temperatures (in the SmC* and the SmC$_A$* phases) versus measuring frequency f for the enantiomer (10 V DC field applied).

Figures 8 and 9 show the imaginary part $\varepsilon''_\perp$ of permittivity for several temperatures in both the SmC* and SmC$_A$* phases (Figure 8) and in the SmC$_A$* phase (Figure 9). In Figure 8, the residual Goldstone mode (G) is seen in the 97–90 °C temperature range. The P_L and P_H modes start to be detectable at 99 °C and 96 °C, respectively. Again, those modes typical for the SmC$_A$* phase coexist with the Goldstone mode typical for the SmC* phase. Please notice that for the temperature range 94–97 °C, there is a clear jump in the strength of the P_L mode, while for the 94–96 °C range, there is a clear jump in the strength of the P_H mode. Figure 9 shows the results for the SmC$_A$* phase. Only the curve for 100 °C does not show any evidence of the P_L or P_H modes. All three modes (P_L, P_H, and S) are well-separated in the frequency domain. At 60 °C, the S mode relaxation frequency is higher than 10 MHz, the P_H mode is around 355 kHz, and the P_L mode is around 630 Hz. When we compare Figures 6 and 9, it is seen that the P_H and P_L modes are a little stronger under the DC field while the S mode is a little suppressed. This effect was reported in many papers [20–22,29,38–41]. When the DC field is applied, the ion contribution (at low frequencies) is reduced, and one can see that the amplitude of the P_L mode increases when the temperature decreases (it is not clear in Figure 6). Additionally, in the SmC_A* phase under the DC field, the residual Goldstone mode is observed at frequencies below the P_L mode (for temperatures of 95–80 °C).

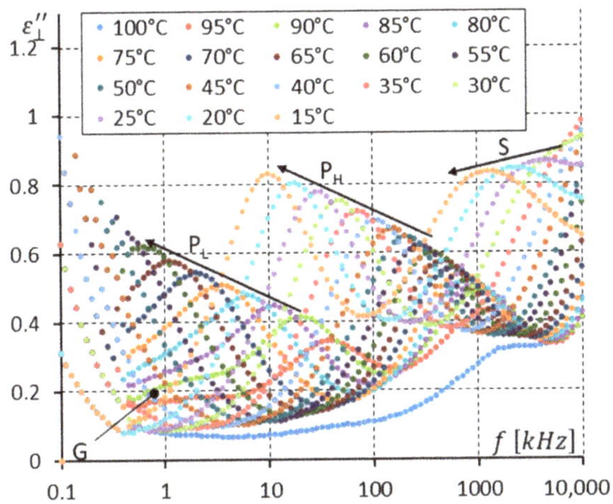

Figure 9. The imaginary part $\varepsilon''_\perp$ of permittivity for eighteen temperatures (in the SmC_A* phase and one temperature (100 °C) in the SmC* phase) versus measuring frequency f for the enantiomer (10 V DC field applied). Arrows show how the maximum amplitude of $\varepsilon''_\perp$ (also the relaxation frequencies) shifts when the temperature decreases. The residual Goldstone mode (G) is also seen.

3.3. Results of the Measurements Performed for the Racemic Mixture

After measuring the enantiomer, measurements of the racemic mixture were performed. Figure 10 shows the real part $\varepsilon'_\perp$ of permittivity versus temperature for twelve frequencies. Only two modes in the SmC_A phase are observed, which can be interpreted as the P_L and S modes. The racemate does not exhibit the third mode observed in the SmC_A* phase—P_H. The mode visible in the racemic SmC phase, denoted as "X", seems to continue the P_L mode observed in the SmC_A phase. The dispersion of the X mode is weaker than that of the P_L mode.

To analyze the dielectric response more precisely, Figures 11 and 12 were prepared. Figure 11 shows the imaginary part $\varepsilon''_\perp$ of permittivity versus frequency for several temperatures (from 100 to 77 °C) at the Iso-SmC phase transition (100, 99 °C) and in the SmC phase and the SmC_A phase (78 and 77 °C). For three temperatures, i.e., 100, 99, and 98 °C, high-frequency relaxation can be interpreted as the molecular motions around the short molecular axis (S mode). The clearing temperature is around 99 °C. This means the S mode shown in Figure 11 corresponds to molecular motion in an isotropic liquid. For temperatures of 97–79 °C, the SmC phase exists, and only one mode is seen. It is denoted as the X mode because it is the first time this mode is analyzed in the synclinic racemic phase. In the SmC phase, the S mode seems to be invisible. When the SmC transforms (while cooling) into the SmC_A phase, the X mode does not disappear. Moreover, its strength is doubled. The low-frequency limit of this mode is constant at the temperature of phase transition, while the high-frequency limit is reduced clearly (Figure 10) when entering the SmC_A phase.

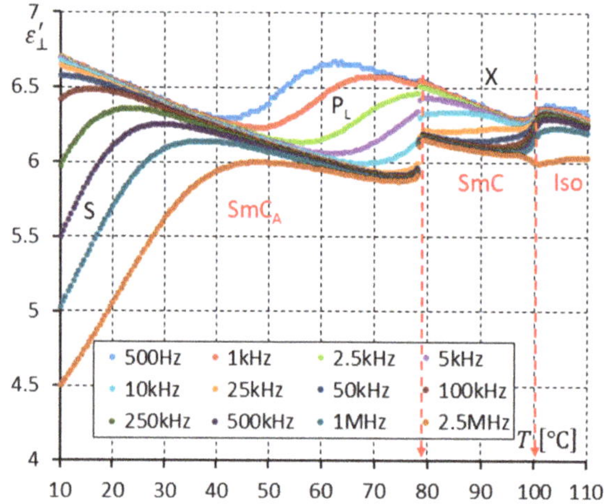

Figure 10. The real part $\varepsilon'_{\perp}$ of permittivity for twelve measuring frequencies versus temperature T for the racemate (no DC field applied). Two modes (P_L, S) are seen in the SmC_A phase, and the X mode is observed in the SmC phase. Arrows indicate the phase transitions.

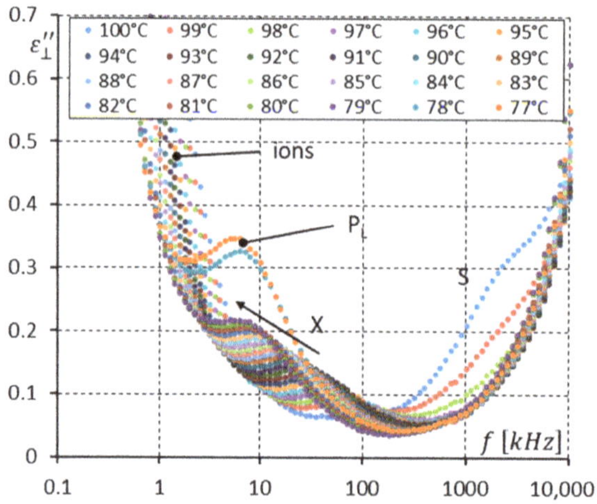

Figure 11. The imaginary part $\varepsilon''_{\perp}$ of permittivity for twenty-four temperatures (in the isotropic liquid close to the Iso-SmC phase transition, the SmC phase, and the SmC_A phase close to the SmC-SmC_A phase transition) versus measuring frequency f for the racemate (no DC field applied). Arrows show how the maximum amplitude of $\varepsilon''_{\perp}$ (also the relaxation frequency) shifts when the temperature decreases. Ions are seen at low frequencies.

Figure 12 presents the imaginary part $\varepsilon''_{\perp}$ of the dielectric response only in the SmC_A phase (temperatures of 78–10 °C). Two modes are seen (P_L and S). Please note that the strength of the S mode without a DC field applied in the enantiomer (Figure 6) and racemate (Figure 12) for low temperatures of ~20 °C is practically the same. This is unsurprising because the molecular motions around the short molecular axis should not depend on chirality.

One can see in Figures 11 and 12 that the imaginary part $\varepsilon''_{\perp}$ of permittivity at low frequency is influenced by the presence of ions. The DC field (10 V) was applied to suppress the ion contribution to the dielectric response. Additionally, it helped to observe how the

DC field modifies the dielectric response in the racemate. The results are presented in Figures 13–15.

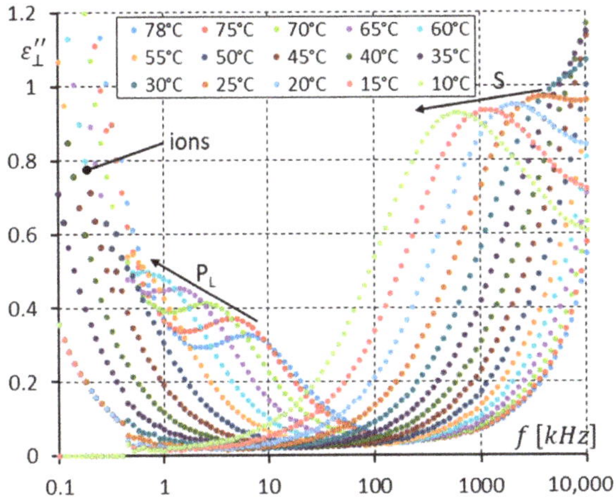

Figure 12. The imaginary part $\varepsilon''_\perp$ of permittivity for fifteen temperatures (in the SmC_A phase) versus measuring frequency f for the racemate (no DC field applied). Arrows show how the maximum amplitude of $\varepsilon''_\perp$ (also the relaxation frequencies) shifts when the temperature decreases. Ions are seen at low frequencies.

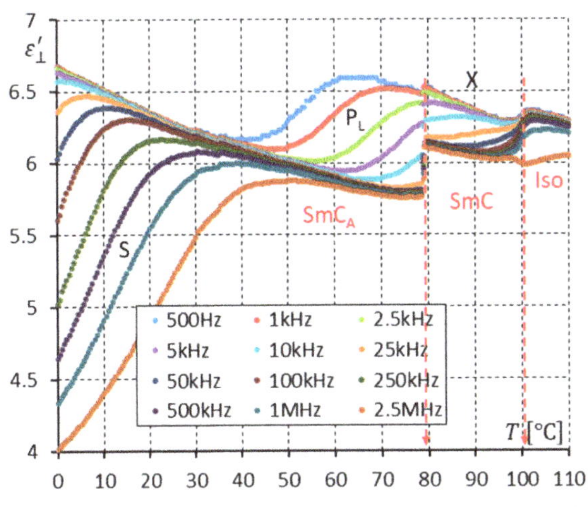

Figure 13. The real part $\varepsilon'_\perp$ of permittivity for twelve measuring frequencies versus temperature T for the racemate (10 V DC field applied). Two modes (P_L, S) are seen in the SmC_A phase, while the X mode is observed in the SmC phase. Arrows indicate the phase transitions.

Figures 10 and 13 show the real part $\varepsilon'_\perp$ of permittivity. They seem to be very similar. One can only find that the high-frequency limits of permittivity in the SmC_A and SmC phases close to the SmC-SmC_A phase transition are lower for permittivity measured with the DC field (Figure 13) than without the DC field (Figure 10).

Essential differences can be seen when comparing Figures 11 and 14, where the imaginary parts $\varepsilon''_\perp$ of permittivity in the Iso, SmC, and SmC_A phases are presented (without and with the DC field, respectively). Because of the effective suppression of the ion contribution, the imaginary part $\varepsilon''_\perp$ of permittivity is well visible in Figure 14 at low frequencies. One can notice that the P_L mode is strengthened under a DC field. Similarly, the X mode in the SmC phase is strengthened under a DC field. The X mode behaves like the P_L mode in the SmC_A phase.

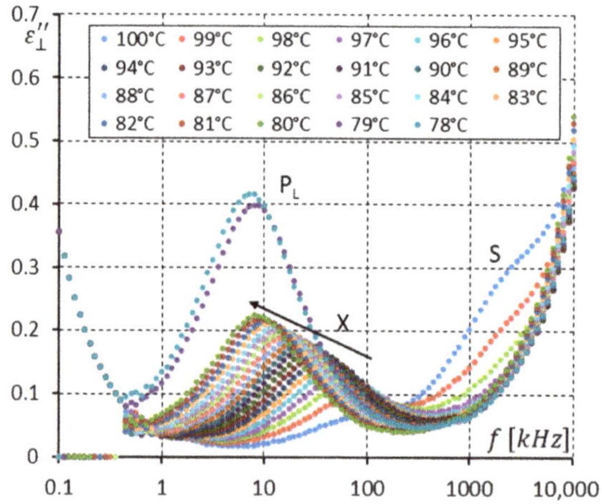

Figure 14. The imaginary part $\varepsilon''_{\perp}$ of permittivity for twenty-three temperatures (in the isotropic liquid close to the Iso-SmC phase transition, the SmC, and the SmC_A close to the SmC-SmC_A phase transition) versus measuring frequency f for the racemate (10 V DC field applied). The arrow shows how the maximum amplitude of $\varepsilon''_{\perp}$ (also the relaxation frequency) shifts when the temperature decreases.

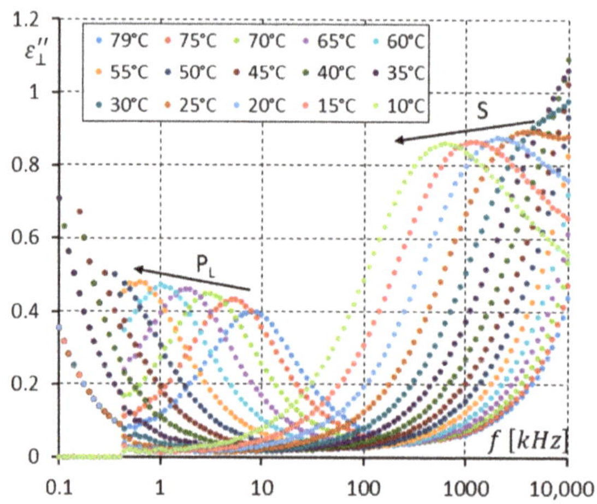

Figure 15. The imaginary part $\varepsilon''_{\perp}$ of permittivity for fifteen temperatures (in the SmC_A phase) versus measuring frequency f for the racemate (10 V DC field applied). Arrows show how the maximum amplitude of $\varepsilon''_{\perp}$ (also the relaxation frequencies) shifts when the temperature decreases.

When comparing Figures 12 and 15, where the imaginary part $\varepsilon''_{\perp}$ of permittivity in the SmC_A phase is presented (without and with the DC field, respectively), the P_L mode is stronger under the DC field, while the S mode is weaker under the DC field.

Three-dimensional plots of the imaginary part $\varepsilon''_{\perp}$ of permittivity versus frequency and temperature are presented in the Supplementary Materials.

3.4. Results of the Calculations

The parameters describing the modes presented in the dielectric response were determined to fit calculated permittivity using Formula (1) with the experimental results. For the racemate, the results of the calculations for temperatures close to the SmC-SmC$_A$ phase transition are presented in Figure 16 (measurements without a DC field) and Figure 17 (measurements under a 10 V DC field). Both figures present the dielectric strengths of the detected modes (P$_L$ and X) and relaxation frequencies of the detected modes (P$_L$ and X). One can notice that the phase transition SmC-SmC$_A$ is seen only in Figures 16a and 17a. The dielectric strength plot is not continuous at the SmC-SmC$_A$ phase transition.

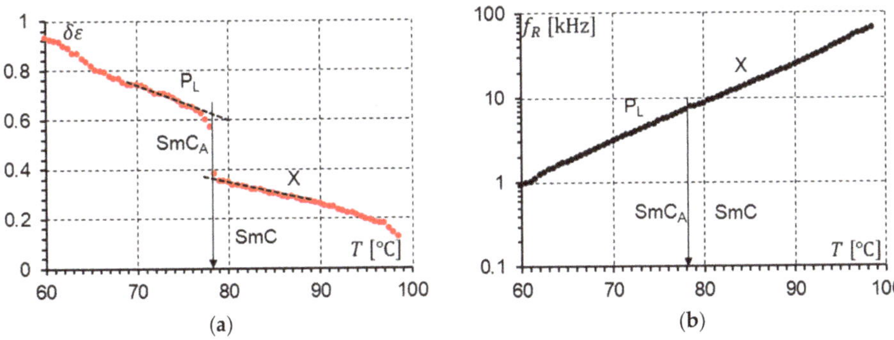

Figure 16. (**a**) The strengths $\delta\varepsilon$ of the P$_L$ and X modes versus temperature at the SmC$_A$-SmC phase transition in the racemate. No DC field was applied. The phase transition is marked by an arrow. (**b**) The relaxation frequency f_R of the P$_L$ and X modes versus temperature at the SmC$_A$-SmC phase transition in the racemate. No DC field was applied. The phase transition is marked by an arrow. The dashed line in the $\delta\varepsilon(T)$ plot is the linear approximation of $\delta\varepsilon(T)$ near the phase transition.

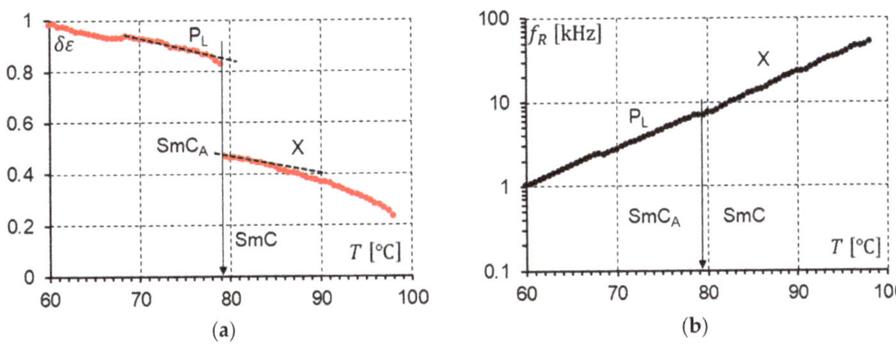

Figure 17. (**a**) The strengths $\delta\varepsilon$ of the P$_L$ and X modes versus temperature at the SmC$_A$-SmC phase transition in the racemate. A 10 V DC field was applied. The phase transition is marked by an arrow. (**b**) The relaxation frequency f_R of the P$_L$ and X modes versus temperature at the SmC$_A$-SmC phase transition in the racemate. A 10 V DC field was applied. The phase transition is marked by an arrow. The dashed line in the $\delta\varepsilon(T)$ plot is the linear approximation of $\delta\varepsilon(T)$ near the phase transition.

When the plot of the relaxation frequency f_R versus temperature T (Figures 16b and 17b) is considered, no phase transition can be seen. The relaxation frequency changes continuously. When the DC field is on, the dielectric strength of the P$_L$ and X modes increases. The X mode shows a similar behavior as the P$_L$ mode. It is worth highlighting that at cooling, when the SmC phase transforms into the SmC$_A$ phase, the dielectric strength increases almost twice (0.63/0.36 = 1.75—without DC field; 0.85/0.47 = 1.81—with 10 V DC field). This value would probably be closer to two when the structure is perfect. This is probably related to the fact that in the synclinic SmC phase, the unit structure consists of one layer, while in the anticlinic SmC_A phase, it consists of two layers. Hence, a doubled structure gives doubled strength.

Knowing the temperature dependence of the relaxation frequencies of the X and P_L modes shown in Figures 16b and 17b, we can use the Arrhenius law (Equation (2)) to determine the activation energy E_a. This is possible when the observed modes fulfill the Arrhenius model. Finally, the activation energy E_a was calculated. The plots prepared for the measurements without (Figure 18a) and with the 10 V DC field (Figure 18b) give almost the same results for activation energy as follows: E_a = 1.14 eV/molecule (Figure 18a) and E_a = 1.10 eV/molecule (Figure 18b). The activation energy is the same for both modes (P_L and X).

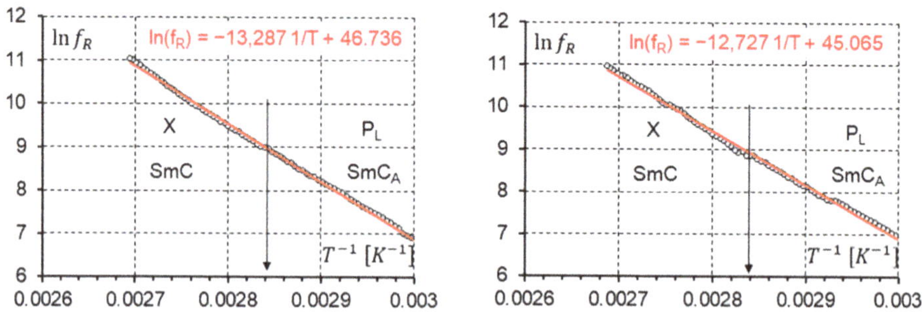

(a) (b)

Figure 18. The Arrhenius plot for the P_L and X modes at the SmC_A-SmC phase transition in the racemate. The phase transition is marked by an arrow. (**a**) No DC field is applied. (**b**) A 10 V DC field is applied. The red line is the linear approximation of $\ln f_R(T^{-1})$.

When the similarities between the P_L (in the SmC_A phase) and X (in the SmC phase) modes are shown, we can calculate the parameters of the modes observed in the racemic SmC_A and the enantiomeric SmC_A*. Below, we show only the results obtained for racemate and enantiomer under the influence of the 10 V DC field because the results measured without a DC field are very similar. Only amplitudes of observed modes in the anticlinic phases (SmC_A and SmC_A*) change under the DC field. Additionally, the DC field reduces the ion contribution to the dielectric response, making it easier to find the parameters of the P_L mode.

Firstly, it must be mentioned that the quality of the relaxation parameters determination depends on the measuring range of the used equipment. In our case, the parameters of the P_L mode can be well determined if its relaxation frequency is higher than 200 Hz, while the parameters of the S mode can be well determined if its relaxation frequency is lower than 10 MHz. This is guaranteed by the method presented in [35]. The parameters of the P_H mode are calculated more precisely because its relaxation frequency is in the middle of the measuring range of our impedance analyzer.

When comparing Figures 19b and 20b, it is seen that the relaxation frequencies of the S and P_L modes in the SmC_A (racemate) and SmC_A* (enantiomer) phases are the same for measurements performed with a 10 V DC field. In SmC_A*, additionally, the P_H mode is observed with an amplitude similar to the P_L mode (Figure 20a). The P_H mode coexists with the S and P_L modes in different temperature ranges. We cannot simultaneously see (similar to the racemic SmC_A) the S and P_L modes at the same temperature.

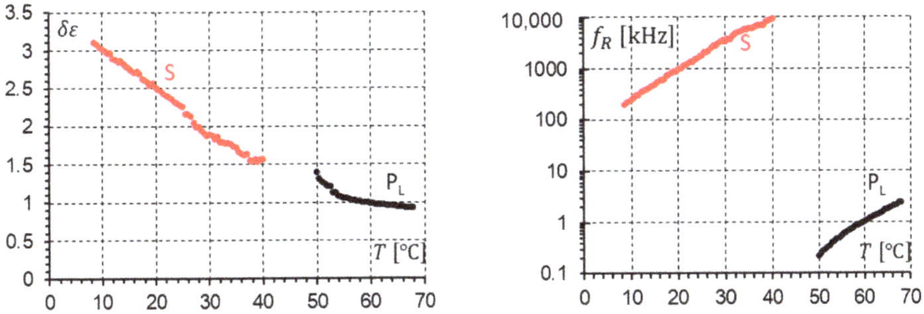

(**a**) (**b**)

Figure 19. (**a**) The strengths $\delta\varepsilon$ of the P_L and S modes versus temperature in the SmC_A phase in the racemate. (**b**) The relaxation frequency f_R of the P_L and S modes versus temperature in the SmC_A phase transition in the racemate. A 10 V DC field was applied.

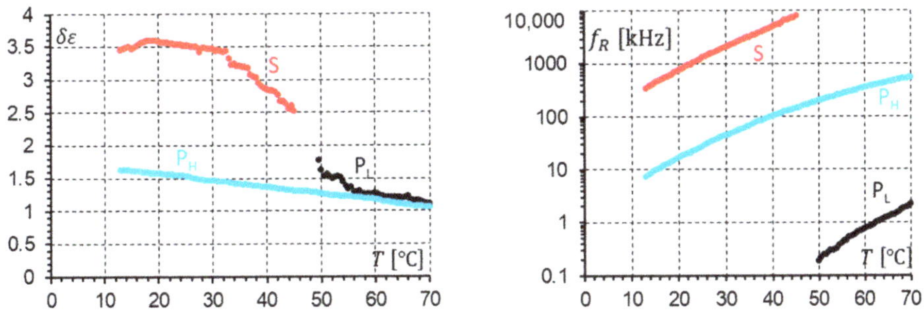

(**a**) (**b**)

Figure 20. (**a**) The strengths $\delta\varepsilon$ of the P_L, P_H, and S modes versus temperature in the SmC_A* phase in the enantiomer. (**b**) The relaxation frequency f_R of the P_L, P_H, and S modes versus temperature in the SmC_A* phase in the enantiomer. A 10 V DC field was applied.

4. Discussion, Conclusions, and Summary

Previous dielectric studies were limited to pure antiferroelectric and ferroelectric enantiomers. Only three papers were devoted to dielectric studies of the "antiferroelectric" racemates [17–19], while the "ferroelectric" racemate did not raise interest so far. The dielectric response of the SmC_A* phase is well described in many papers [20–24,38–44]. Four modes in the antiferroelectric (SmC_A*) phase were found.

Two molecular modes are related to individual motions of molecules around a long molecular axis (L mode) and a short molecular axis (S mode). The L mode's relaxation frequency is higher than the measuring range of our experimental setup. Hence, only the S mode was detected in the experiment shown here. The sample should be cooled enough (below 30 °C) to make this mode visible in measurements. The S mode and the L mode are Arrhenius-type modes. This means that their relaxation frequencies decrease with a decrease in temperature. To observe the L mode, the measurement temperature should be below −50 °C [45]. In the presented experiment, the lowest temperature was higher than 0 °C.

The next two modes (P_L and P_H) were described as "non-cancelled phason modes" [20,21,23]. Their amplitude is lower than the amplitude of the S mode. This means that they are weak.

For the same temperature, the relaxation frequency of the P_H mode is 300–500 times higher than the relaxation of the P_L mode. One may think that the P_L mode seems similar to the Goldstone mode in the ferroelectric SmC* phase, but our results show that the dielectric responses of the Goldstone mode and the P_L mode at the phase transition exhibit different relaxation frequencies. Our results show that Goldstone and P_L modes can coexist at temperatures close to the SmC*-SmC_A* phase transition. Both the P_H and P_L modes are stronger under a low DC field—this effect is also presented in this paper. However, when we apply a high enough DC field, the helical structure becomes unwound, and both the P_H and P_L modes disappear [29]. This means that unwound (by a DC field) helical structures suppress the Goldstone, P_L, and P_H modes. Moreover, this paper confirms that unwinding the helical structure in the SmC_A* phase is more difficult than in the SmC* phase.

The racemization changes the dielectric response in comparison with the pure enantiomer. One can observe several effects, which can be found using dielectric spectroscopy. Racemization fully suppresses the P_H mode in the SmC_A phase (Figures 10, 12, 13 and 15). This mode is visible in the enantiomer only (Figures 4, 6, 7 and 9). Simultaneously in the racemate, the P_L and S modes in the SmC_A phase are still detectable as they are in the enantiomeric SmC_A* phase (compare Figures 4 and 10 and compare Figures 6 and 12). The P_L mode in the racemic SmC_A phase is amplified under the DC field (compare Figures 10 and 13 and compare Figures 12 and 15), similar to that in the enantiomeric SmC_A* phase (compare Figures 4 and 7 and compare Figures 6 and 9). Simultaneously, the S mode in the racemic SmC_A phase is attenuated under the DC field (Figures 12 and 15) as it is in the enantiomeric SmC_A* phase (Figures 6 and 9). The relaxation frequencies of the P_L and S modes in the racemic SmC_A phase (Figure 19b) are approximately the same as the relaxation frequencies of the P_L and S modes in the enantiomeric SmC_A* phase (Figure 20b). Moreover, the S mode is stronger than the P_L mode in the racemic SmC_A phase (Figure 19a) and in the enantiomeric SmC_A* phase (Figure 20a). The X mode found in the racemic SmC phase seems to continue the P_L mode from the racemic SmC_A phase. Additionally, the amplitude $\delta\varepsilon$ of the P_L mode in the racemic SmC_A phase is almost twice as high as the amplitude of the X mode in the racemic SmC phase close to the SmC-SmC_A phase transition (Figures 16a and 17a). A similar effect is visible at the SmC*-SmC_A* phase transition (Figure 8). The Goldstone mode is suppressed when the 10V DC field is on in the SmC* phase, and weak P_L and P_H modes are well seen. A similar effect when racemization reveals modes, normally covered by a strong Goldstone mode, was observed earlier [46]. The P_L mode is amplified by a factor of two when the SmC* changes into the SmC_A* phase (compare plots for temperatures of 97 and 95 °C). The relaxation frequency f_R of the X mode in the racemic SmC phase decreases with as the temperature decreases, and the same is observed for the relaxation frequency of the P_L mode in the racemic SmC_A (Figures 16b and 17b). One can see the continuous evolution of the relaxation frequency at the SmC-SmC_A phase transition. The X mode in the racemic SmC phase is amplified under the DC field as it is for the P_L mode in the racemic SmC_A phase (Figures 11 and 14). One should remember that ions can distort the low-frequency imaginary part of complex permittivity—the DC field suppresses this effect efficiently. The Goldstone mode is not detectable in the racemic SmC phase (compare Figures 4 and 10 and compare Figures 5 and 11). The Goldstone mode in the enantiomeric SmC* phase exhibits a lower relaxation frequency than the P_L mode in the enantiomeric SmC_A* phase (Figures 5 and 8). These two modes are different—they have nothing to do with each other.

One should remember that molecules in the racemate are still chiral. But the equimolar R- and S-enantiomer ratio makes the mixture optically inactive (the helicoidal structure disappears). Racemization is the procedure that should suppress or eliminate all material properties related to chirality. This is obviously seen in the disappearance of the Goldstone mode in the racemic SmC phase. It is still unclear why only the P_H mode is suppressed in SmC_A by racemization while the P_L mode seems to be not influenced by racemization; a theoretical explanation is required. This experimental fact should change the opinion that the P_L and P_H modes have similar origins in chirality [20,21,38,39]. The fact that both the P_L and P_H modes exhibit very similar behavior in the enantiomer does not rule out the possibility that both modes will behave completely differently in the racemate.

Moreover, the racemization in the racemic SmC phase reveals the X mode, similar to the P_L mode in the racemic SmC_A and the enantiomeric SmC_A* phases. A similar continuous transformation of the mode observed in the SmC and SmC_A phases was observed in an achiral compound [47]. When racemization eliminates the strong Goldstone mode, a weak X mode (P_L mode-like) can be visible. When we compare the results from Figures 8 and 14, we can venture to say that the P_L and P_H modes start to be seen in the SmC* phase (close to the SmC*-SmC_A* phase transition) when the Goldstone mode is suppressed by the DC field. Similarly, the residual Goldstone mode is still detected in the SmC_A* phase (close to the SmC*-SmC_A* phase transition).

The conclusions about dielectric modes are collected in Table 1 (for enantiomer) and Table 2 (for racemate).

Table 1. Dielectric modes observed in enantiomeric anticlinic SmC_A* and synclinic SmC* phases when the DC field is off and on.

	SmC_A*	SmC*
Dielectric modes (no DC)	Residual Goldstone mode P_L mode P_H mode Molecular S mode	Goldstone mode Weak P_L mode covered by the Goldstone mode Weak P_H mode covered by the Goldstone mode Molecular S mode
Dielectric modes (with DC)	Residual Goldstone mode P_L mode strengthened by the DC field P_H mode strengthened by the DC field Molecular S mode weakened by the DC field	Suppressed Goldstone mode P_L mode strengthened by the DC field, non-covered by the suppressed Goldstone mode P_H mode strengthened by the DC field, non-covered by the suppressed Goldstone mode Molecular S mode weakened by the DC field

Table 2. Dielectric modes observed in racemic anticlinic SmC_A and synclinic SmC phases when the DC field is off and on.

	SmC_A	SmC
Dielectric modes (no DC)	P_L mode Molecular S mode	Weak X(P_L) mode Molecular S mode
Dielectric modes (with DC)	P_L mode strengthened by the DC field Molecular S mode weakened by the DC field P_L mode strengthened by the DC field P_H mode strengthened by the DC field Molecular S mode weakened by the DC field	X(P_L) mode strengthened by the DC field Molecular S mode weakened by the DC field P_H mode strengthened by the DC field, non-covered by the suppressed Goldstone mode Molecular S mode weakened by the DC field

Supplementary Materials: The following supporting information can be downloaded at: https://www.mdpi.com/article/10.3390/ma17133335/s1. Figure S1. DSC plot for enantiomer. Figure S2. DSC plot for racemate. Figure S3. 3D plot of the imaginary part $\varepsilon_\perp''$ of permittivity versus temperature T [°C] and frequency f [kHz] for the enantiomer (10 V DC field applied). The SmC* phase under 10 V DC field does not show any relaxation. Figure S4. 3D plot of the imaginary part $\varepsilon_\perp''$ of permittivity versus temperature T [°C] and frequency f [kHz] for the racemate (10 V DC field applied). Figure S5. 3D plot of the imaginary part $\varepsilon_\perp''$ of permittivity versus temperature T [°C] and frequency f [kHz] for the enantiomer (5 V DC field applied). The SmC* phase under 5 V DC field shows residual Goldstone mode. Figure S6. 3D plot of the imaginary part $\varepsilon_\perp''$ of permittivity versus temperature T [°C] and frequency f [kHz] for the racemate (5 V DC field applied). Figure S7. 3D plot of the imaginary part $\varepsilon_\perp''$ of permittivity versus temperature T [°C] and frequency f [kHz] for the enantiomer (no DC field applied). The SmC* phase without the DC field shows strong Goldstone mode. Figure S8. 3D plot of the imaginary part $\varepsilon_{\perp}''$ of permittivity versus temperature T [°C] and frequency f [kHz] for the enantiomer (no DC field applied). The SmC* phase without the DC field shows strong Goldstone mode. Figure S9. 3D plot of the imaginary part $\varepsilon_{\perp}''$ of permittivity versus temperature T [°C] and frequency f [kHz] for the racemate (no DC field applied).

Author Contributions: Conceptualization, P.P. and M.U.; methodology, P.P. and M.U.; software, P.P.; validation, P.P. and M.U.; formal analysis, P.P. and M.U.; investigation, P.P. and M.U.; resources, P.P. and M.U.; data curation, P.P. and M.U.; writing—original draft preparation, P.P. and M.U.; writing—review and editing, P.P. and M.U. All authors have read and agreed to the published version of the manuscript.

Funding: This research was funded by a University Research Grant (by the Military University of Technology) no. UGB 22-804 in 2023.

Institutional Review Board Statement: Not applicable.

Informed Consent Statement: Not applicable.

Data Availability Statement: The data presented in this study are available on request from the corresponding author (due to privacy).

Conflicts of Interest: The authors declare no conflicts of interest.

References

1. Thomson, W. *Baltimore Lectures on Molecular Dynamics and the Wave Theory of Light*; Cambridge Library Collection—Physical Sciences, Cambridge University Press: Cambridge, UK, 2010.
2. Mandle, R.J.; Cowling, S.J.; Goodby, J.W. A nematic to nematic transformation exhibited by a rod-like liquid crystal. *Phys. Chem. Chem. Phys.* **2017**, *19*, 11429–11435. [CrossRef] [PubMed]
3. Nishikawa, H.; Shiroshita, K.; Higuchi, H.; Okumura, Y.; Haseba, Y.; Yamamoto, S.-I.; Sago, K.; Kikuchi, H. A fluid liquid-crystal material with highly polar order. *Adv. Mater.* **2017**, *29*, 1702354. [CrossRef] [PubMed]
4. Mandle, R.J.; Cowling, S.J.; Goodby, J.W. Rational design of rod-like liquid crystals exhibiting two nematic phases. *Chemistry* **2017**, *23*, 14554–14562. [CrossRef] [PubMed]
5. Meyer, R.B. Piezoelectric effects in liquid crystals. *Phys. Rev. Lett.* **1969**, *22*, 918–921. [CrossRef]
6. Meyer, R.B.; Liebert, L.; Strzelecki, L.; Keller, P. Ferroelectric liquid crystals. *J. Phys. Lett.* **1975**, *36*, 69–71. [CrossRef]
7. Buka, A.; Stegemeyer, H. Molecular rotational potentials in the S_C* phase of binary ferroelectric liquid crystal mixtures. *Liq. Cryst.* **1990**, *8*, 229–235. [CrossRef]
8. Stegemeyer, H.; Meister, R.; Altenbach, H.J.; Szewczyk, D. Ferroelectricity of induced S^*_C phases with novel chiral dopants. *Liq. Cryst.* **1993**, *14*, 1007–1019. [CrossRef]
9. Stegemeyer, H.; Meister, R.; Hoffmann, U.; Sprick, A.; Becker, A. Ferroelectricity of induced smectic C* phases. *J. Mater. Chem.* **1995**, *5*, 2183–2193. [CrossRef]
10. Raszewski, Z.; Rutkowska, J.; Kędzierski, J.; Perkowski, P.; Piecek, W.; Zieliński, J.; Żmija, J.; Dąbrowski, R. Investigation of the biased rotation of chiral molecule around its long molecular axis in the ferroelectric liquid crystal mixture. *Mol. Cryst. Liq. Cryst.* **1997**, *302*, 85–91. [CrossRef]
11. Reinitzer, F. Contributions to the understanding of cholesterol. *Monatshefte Chem.* **1888**, *9*, 421–441. [CrossRef]
12. Hoffmann, J.; Kuczyński, W.; Małecki, J. Dielectric study of ferroelectric properties in chiral smectic C. *Mol. Cryst. Liq. Cryst.* **1978**, *44*, 287–300. [CrossRef]
13. Chandani, A.D.L.; Górecka, E.; Ouchi, Y.; Takezoe, H.; Fukuda, A. Antiferroelectric chiral smectic phases responsible for the tristable switching in MHPOBC. *Jpn. J. Appl. Phys.* **1989**, *29*, L1265–L1268. [CrossRef]
14. Lavrentovich, O. Ferroelectric nematic liquid crystal, a century in waiting. *Proc. Natl. Acad. Sci. USA* **2020**, *117*, 14629–14631. [CrossRef] [PubMed]
15. Muševič, I.; Blinc, R.; Žekš, B. *The Physics of Ferroelectric and Antiferroelectric Liquid Crystals*; World Scientific Publishing: Singaspore, 2000.
16. Filipič, C.; Carlsson, T.; Levstik, A.; Žekš, B.; Blinc, R.; Gouda, F.; Lagerwall, S.T.; Skarp, K. Dielectric properties near the smectic-C*—Smectic-A phase transition of some ferroelectric liquid-crystalline systems with a very large spontaneous polarization. *Phys. Rev. A* **1988**, *38*, 5833–5839. [CrossRef] [PubMed]
17. Kumari, S.; Das, I.M.L.; Dhar, R.; Dąbrowski, R. Thermodynamic and dielectric spectroscopy of a racemic antiferroelectric liquid crystal. *J. Mol. Liq.* **2012**, *168*, 54–60. [CrossRef]
18. Perkowski, P.; Ogrodnik, K.; Żurowska, M.; Piecek, W.; Dąbrowski, R.; Raszewski, Z.; Jaroszewicz, L. Antiferroelectric and ferroelectric SmC* phases in racemic mixture. *Phase Transit.* **2013**, *86*, 138–146. [CrossRef]

19. Verma, R.; Dąbrowski, R.; Żurowska, M.; Dhar, R. Enhancement of the properties and mesophases stability after the electron beam irradiation on a racemic anti-ferroelectric liquid crystalline mixture. *Liq. Cryst.* **2016**, *43*, 606–614. [CrossRef]
20. Buivydas, M.; Gouda, F.; Lagerwall, S.T.; Stebler, B. The molecular aspect of the double absorption peak in the dielectric spectrum of the antiferroelectric liquid crystal phase. *Liq. Cryst.* **1995**, *18*, 879–886. [CrossRef]
21. Buivydas, M.; Gouda, F.; Andersson, G.; Lagerwall, S.T.; Stebler, B.; Bömelburg, J.; Heppke, G.; Gestblom, B. Collective and non-collective excitations in antiferroelectric and ferrielectric liquid crystals studied by dielectric relaxation spectroscopy and electro-optic measurements. *Liq. Cryst.* **1997**, *23*, 723–739. [CrossRef]
22. Deptuch, A.; Lalik, S.; Jasiurkowska-Delaporte, M.; Juszyńska-Gałązka, E.; Drzewicz, A.; Urbańska, M.; Marzec, M. Comparative study of electrooptic, dielectric, and structural properties of two glassforming antiferroelectric mixtures with a high tilt angle. *Phys. Rev. E* **2022**, *105*, 024705. [CrossRef]
23. Panarin, Y.P.; Kalinovskaya, O.; Vij, J.K. The investigation of the relaxation processes in antiferroelectric liquid crystals by broad band dielectric and electro-optic spectroscopy. *Liq. Cryst.* **1998**, *25*, 241–252. [CrossRef]
24. Marzec, M.; Mikułko, A.; Wróbel, S.; Haase, W.; Dąbrowski, R. Antiferroelectric modes behaviour in a wide temperature range $SmC_A{}^*$ phase. *Mol. Cryst. Liq. Cryst.* **2005**, *437*, 169–180. [CrossRef]
25. Różański, S.; Thoen, J. Collective and antiferroelectric dielectric modes in a highly tilted three-ring ester. *Liq. Cryst.* **2007**, *34*, 519–526. [CrossRef]
26. Mishra, A.; Dąbrowski, R.; Dhar, R. Dielectric characteristics of highly ionic antiferroelectric liquid crystalline material. *J. Mol. Liq.* **2018**, *249*, 106–109. [CrossRef]
27. Perkowski, P.; Raszewski, Z.; Piecek, W.; Ogrodnik, K.; Żurowska, M.; Dąbrowski, R.; Sun, X.-W. Dielectric Properties of New Fluorinated Orthoconic Antiferroelectric Liquid Crystals. *Mol. Cryst. Liq. Cryst.* **2009**, *509*, 328–335. [CrossRef]
28. Huang, L.C.; Fu, C.-M. Spontaneous polarization and dielectric relaxation dynamics of ferroelectric liquid crystals derived from 2(S)-[2(S)-ethylhexyolxy] propionic acid and its (S, R)-diastereomer. *Phys. B* **2015**, *473*, 54–61. [CrossRef]
29. Perkowski, P.; Ogrodnik, K.; Piecek, W.; Żurowska, M.; Raszewski, Z.; Dąbrowski, R.; Jaroszewicz, L. Influence of the bias field on dielectric properties of the $SmC_A{}^*$ in the vicinity of the SmC*-$SmC_A{}^*$ phase transition. *Liq. Cryst.* **2011**, *38*, 1159–1167. [CrossRef]
30. Parab, S.S.; Malik, M.K.; Dąbrowski, R.; Deshmukh, R.R. Thermodynamic and bias field characterization of a quickly operating antiferroelectric liquid crystal. *J. Mol. Liq.* **2013**, *183*, 20–25. [CrossRef]
31. Żurowska, M.; Dąbrowski, R.; Dziaduszek, J.; Garbat, K.; Filipowicz, M.; Tykarska, M.; Rejmer, W.; Czupryński, K.; Spadło, A.; Bennis, N.; et al. Influence of alkoxy chain length and fluorosubstitution on mesogenic and spectral properties of high tilted antiferroelectric esters. *J. Mater. Chem.* **2011**, *21*, 2144–2153. [CrossRef]
32. Vojtylová, T.; Żurowska, M.; Milewska, K.; Hamplová, V.; Sýkora, D. Chiral HPLC and physical characterisation of orthoconic antiferroelectric liquid crystals. *Liq. Cryst.* **2016**, *43*, 1244–1250. [CrossRef]
33. Perkowski, P. The parasitic effects in high-frequency dielectric spectroscopy of liquid crystals—The review. *Liq. Cryst.* **2021**, *48*, 767–793. [CrossRef]
34. Dhar, R. An impedance model to improve the higher frequency limit of electrical measurements on the capacitor cell made from electrodes of finite resistances. *Int. J. Pure Appl. Phys.* **2004**, *42*, 56–61. [CrossRef]
35. Perkowski, P.; Mrukiewicz, M.; Laska, M.; Garbat, K.; Piecek, W.; Dąbrowski, R. Dielectric behavior of dual-frequency nematic at extra low temperatures. *Phase Transit.* **2013**, *86*, 113–122. [CrossRef]
36. Havriliak, S.; Negami, S. A Complex plane representation of dielectric and mechanical relaxation processes in some polymers. *Polymer* **1967**, *8*, 161–210. [CrossRef]
37. Bötcher, C.J.F.; Bordewijk, P. *Theory of Electric Polarization*; Elsevier: Amsterdam, The Netherlands, 1978.
38. de la Fuente, M.R.; Merino, S.; Gonzalez, Y.; Perez Jubindo, M.A.; Ros, B.; Puertolas, J.A.; Castro, M. Dielectric relaxation processes in an antiferroelectric liquid crystal. *Adv. Mater.* **1995**, *7*, 564–568. [CrossRef]
39. Moritake, H.; Uchiyama, Y.; Myojin, K.; Ozaki, M.; Yoshino, K. Dielectric properties in antiferroelectric liquid crystals and their DC bias effects. *Ferroelectrics* **1993**, *147*, 53–66. [CrossRef]
40. Hiraoka, K.; Takezoe, H.; Fukuda, A. Dielectric relaxation modes in the antiferroelectric smectic $C_A{}^*$ phase. *Ferroelectrics* **1993**, *147*, 13–25. [CrossRef]
41. Czerwiec, J.M.; Dąbrowski, R.; Garbat, K.; Marzec, M.; Tykarska, M.; Wawrzyniak, A.; Wróbel, S. Dielectricand electro-optic behaviour of two chiral com-pounds and their antiferroelectric mixtures. *Liq. Cryst.* **2012**, *39*, 1503–1511. [CrossRef]
42. Carvalho, P.S.; Chaves, M.R.; Destrade, C.; Nguyen, H.T.; Glogarova, M. Ferroelectric and antiferroelectric modes in a new chiral thiobenzoate liquid crystal. *Liq. Cryst.* **1996**, *21*, 31–37. [CrossRef]
43. Fitas, J.; Dłubacz, A.; Fryń, P.; Marzec, M.; Jaworska-Gołąb, T.; Deptuch, A.; Kurp, K.; Tykarska, M.; Żurowska, M. New ferroelectric and antiferroelectric liquid crystals studied by complementary methods. *Liq. Cryst.* **2016**, *44*, 566–576. [CrossRef]
44. Pandey, M.B.; Dąbrowski, R.; Dhar, R. Investigation of relaxation processes in anticlinic smectic C* ($SmC_A{}^*$) phase of Liquid Crystals by dielectric spectroscopy. *Phys. B Cond. Matt.* **2007**, *387*, 25–31. [CrossRef]
45. Perkowski, P.; Mrukiewicz, M.; Żurowska, M.; Dąbrowski, R.; Jaroszewicz, L. Dielectric modes in antiferroelectric liquid crystal observed at low temperatures. *Liq. Cryst.* **2013**, *40*, 864–870. [CrossRef]

46. de la Fuente, M.R.; Martín, E.; Pérez Jubindo, M.A.; Artal, C.; Ros, B.; Serrano, J.L. Achiral diacrylate with an anticlinic smectic phase. *Liq. Cryst.* **2001**, *28*, 151–155. [CrossRef]
47. Palacios, B.; de La Fuente, M.R.; Pérez Jubindo, M.A.; Ros, B. Dielectric study of collective and molecular dynamics in a glass-forming low molecular mass ferroelectric liquid crystal and its racemic mixture. *Liq. Cryst.* **1997**, *23*, 349–355. [CrossRef]

materials

MDPI

Article

Molecular Dynamics, Dielectric Properties, and Textures of Protonated and Selectively Deuterated 4′-Pentyl-4-biphenylcarbonitrile Liquid Crystal

Jadwiga Tritt-Goc [1], Magdalena Knapkiewicz [1], Piotr Harmata [2], Jakub Herman [2] and Michał Bielejewski [1,*]

1 Institute of Molecular Physics, Polish Academy of Sciences, M. Smoluchowskiego 17, 60-179 Poznan, Poland; jtg@ifmpan.poznan.pl (J.T.-G.); magdalena.knapkiewicz@ifmpan.poznan.pl (M.K.)
2 Institute of Chemistry, Military University of Technology, ul. gen. S. Kaliskiego 2, 00-908 Warsaw, Poland; piotr.harmata@wat.edu.pl (P.H.); jakub.herman@wat.edu.pl (J.H.)
* Correspondence: bielejewski@ifmpan.poznan.pl

Abstract: Using liquid crystals in near-infrared applications suffers from effects related to processes like parasitic absorption and high sensitivity to UV-light exposure. One way of managing these disadvantages is to use deuterated systems. The combined ^{1}H and ^{2}H nuclear magnetic resonance relaxometry method (FFC NMR), dielectric spectroscopy (DS), optical microscopy (POM), and differential scanning calorimetry (DSC) approach was applied to investigate the influence of selective deuteration on the molecular dynamics, thermal properties, self-organization, and electric-field responsiveness to a 4′-pentyl-4-biphenylcarbonitrile (5CB) liquid crystal. The NMR relaxation dispersion (NMRD) profiles were analyzed using theoretical models for the description of dynamics processes in different mesophases. Obtained optical textures of selectively deuterated 5CB showed the occurrence of the domain structure close to the I/N phase transition. The dielectric measurements showed a substantial difference in switching fields between fully protonated/deuterated 5CB and selectively deuterated molecules. The DSC thermograms showed a more complex phase transition sequence for partially deuterated 5CB with respect to fully protonated/deuterated molecules.

Keywords: nematic liquid crystal; 5CB series; molecular dynamics; fast field cycling NMR relaxometry; deuteration procedure; LC textures

1. Introduction

Liquid crystals (LCs) are materials that combine order and mobility on a molecular and supramolecular level, displaying physical properties that are dependent on the self-organization and molecular dynamics of LC molecules. The versatile and unique properties of these materials and their mixtures are mostly associated with display technologies. However, the high sensitivity to external physical and chemical environments influencing the self-organization of LC molecules enables them to be used in a variety of other fields, such as LC-based chemical sensors [1], biosensors [2], photovoltaics [3] and nanophotonics [4], light-emitting diodes [5], molecular motors [6], field-effect transistors [7], and lasers [8–10]. The possibility of obtaining bi- and multicomponent mixtures of LC materials and chemical modification of base materials allows for obtaining LC metamaterials with a broad range of modifications, e.g., magnetic or chromonic properties [11].

Although liquid crystals have been known since the late 19th century [12] and since then have been intensively studied and developed, there are still some aspects where improvement needs to be carried out to utilize liquid crystals in new fields. One such emerging area is near-infrared range applications where mesomorphic organic materials exhibit several disadvantages, such as parasitic absorption. This unwanted effect relies on the presence of absorption from molecular vibration bands, their overtones, and combination bands, primarily related to the vibrations of C-H bonds in rings and alkyl chains [13–15].

LCs used in devices operating in the infrared are commonly modified, as an example, we can distinguish dye-doped chiral nematic liquid crystals [16] or polymer-dispersed liquid crystal/graphene oxide nanocomposites [17].

A natural way to reduce absorption in this electromagnetic spectrum range is to treat the molecule as a harmonic oscillator and increase the mass of one of the oscillator's elements, specifically the C-H bond. This adjustment reduces the frequency of its fundamental vibrations and shifts the absorption bands toward longer wavelengths. One of the possible alternatives that can be considered for the C-H bond is to replace the C-H bond with a C-D bond. The fundamental valence vibration frequency of the C-H bond (3030–2778 cm^{-1}, or 3.3–3.6 μm) is shifted to the range of 2272–2040 cm^{-1} (4.4–4.9 μm) for C-D bonds. This shift also affects the first harmonic of the C(aryl)-H bond deformation vibration, moving it from approximately 5882 cm^{-1} (1.7 μm for C-H) to 4348 cm^{-1} (2.3 μm for C-D). The higher harmonics are of minimal importance due to their very low intensity, less than 0.001 times the base intensity.

The idea of using deuterated materials for this purpose has been known since the 1980s [15,18]. However, the complexity of synthesizing deuterated mesogens has effectively prevented a more comprehensive range of materials from being obtained and studied. There are reports in the literature on the synthesis of deuterated liquid crystals, but all are limited to the presentation of the main homolog of the thermotropic liquid crystal family, i.e., 4-cyano-4′-pentylbiphenyl 5CB (also used abbreviated name PCB) [18–21].

Deuterated 5CB liquid crystal materials can be synthesized using two primary methodologies: the hydrogen–deuterium exchange method and direct synthesis from deuterated precursors. The hydrogen–deuterium exchange method involves creating the conventional hydrocarbon skeleton, specifically 4-alkylbiphenyl, which undergoes hydrogen–deuterium exchange using catalysts like palladium or platinum and D_2O as the deuterium source, with the polar CN group introduced at the final step. However, this method often results in incomplete deuterium exchange, especially in the alkyl chain segments, achieving less than 85–90% deuteration. Alternatively, the direct synthesis method starts with deuterated raw chemicals of high purity (over 99% deuterium atoms) and synthesizes the liquid crystal molecule through conventional multistep organic synthesis, ensuring a high level of isotopic purity throughout the molecule. The materials discussed here were synthesized following our modified literature procedure described in ref. [20].

Additionally, to reduce parasitic absorption in the NIR spectral range in deuterated LCs, another promising aspect of substituting the ^{1}H atoms in the structure of the molecules with the ^{2}H isotope has been found. Our latest work on deuterated-labeled liquid crystal materials has shown that such materials exhibit enhanced stability under ultraviolet (UV) radiation compared to their nondeuterated counterparts [22,23]. Compared to hydrogen, the increased mass of deuterium atoms results in stronger C-D bonds that are less prone to photochemical degradation. This enhancement in bond strength reduces the rate of bond cleavage and the formation of reactive intermediates under UV exposure, thereby improving the overall durability and longevity of liquid crystal materials when subjected to UV light. This discovery opens new possibilities for applying deuterated liquid crystals in environments with high UV exposure, extending their functional lifespan and performance. However, to successfully utilize deuterated liquid crystals for applications in a wider spectral range of the electromagnetic field, the influence of selective and controlled isotopic substitution on crucial physical properties must be determined.

In the current study, we chose the 4′-pentyl-4-biphenylcarbonitrile (5CB) liquid crystal as a model system to investigate the influence of selective deuteration on its molecular dynamics, mesophase occurrence, and electrical properties. The 5CB is a representative of widely used material whose properties have been investigated by different experimental techniques, allowing us to determine its properties accurately. However, most experimental techniques deliver only a single and averaged component of some second-rank property like susceptibility, permittivity, or refractive index. Because most of the mesogenic materials stay in dynamic equilibrium at a given LC phase, they involve a number of conformational states that average over the molecular orientations, with intra- and intermolecular interactions fluctuating in time. To investigate the molecular dynamics and its dependence on the self-organization of LC molecules, we used the nuclear magnetic resonance relaxometry method in the function of the external magnetic field to determine the spectral density functions of the investigated systems. To confirm and check how the chemical structure of the LC molecule affects its physical properties, we selectively deuterated 5CB, introducing deuterons either in the core of the molecule (5CB-d_8) or substituting all hydrogen atoms (5CB-d_{19}) and compared the results with the fully protonated 5CB molecule.

The purpose of the current work is to study the influence of controlled deuterization on spin–lattice relaxation mechanisms, phase occurrence, and thermal properties in 5CB liquid crystal in isotropic and nematic phases to assess the information on the effect of deuterization degree on molecular dynamics and self-organization of LC molecules chemically modified for utilization in near-infrared applications. Our aim is to determine if observed differences in nematic phase textures, the presence of additional phase transition peaks, and responsiveness to external electric fields observed for different degrees of deuteration can be related to local dynamic processes. The advantage of the NMR FFC method is the accessible time scale of motions that can be detected in the range from 10^{-8} to 10^{-3} s, being specifically sensitive for slow motions that are pronounced for different mesophases in the low-frequency range.

2. Materials and Methods

The samples of 5CB, 5CB-d_8, and 5CB-d_{19} were synthesized in the LC group of the Military University of Technology, Warsaw, according to the procedure described elsewhere [20]. Table 1 presents the structural formulas of the studied materials.

Table 1. Acronyms, chemical structures, chemical and isotopic purities of the studied materials.

Acronym	Chemical Structure	Chemical Purity	Isotopic Purity
5CB	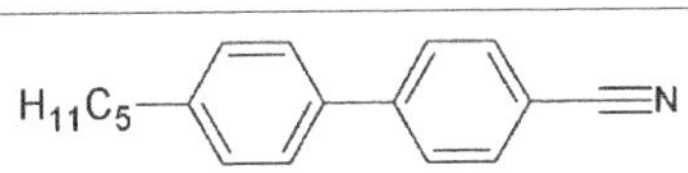	99.8%	^{1}H—100%
5CB-d_8	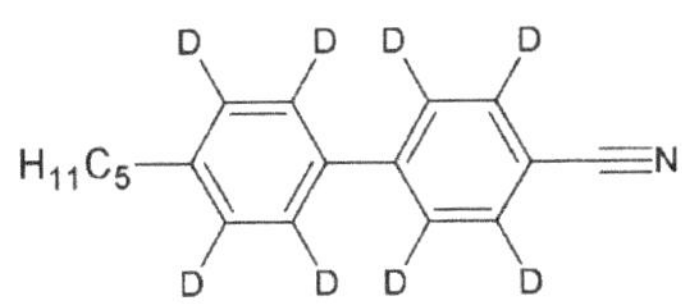	99.4%	^{2}H total—47% ^{2}H rings—96% ^{2}H chain—11%
5CB-d_{19}	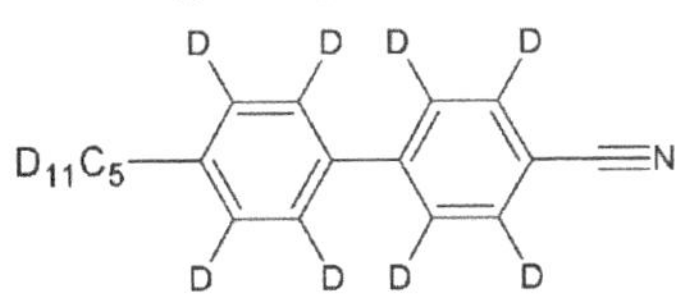	99.5%	^{2}H total—96% ^{2}H rings—93% ^{2}H chain—97%

Chemical purity determined by gas chromatography–mass spectrometry method (GC-MS); isotopic purity determined by ^{1}H nuclear magnetic resonance method (NMR).

2.1. NMR Relaxometry

The molecular dynamics investigations were performed using the nuclear magnetic resonance method and fast field cycling relaxometry technique. The NMR dispersion (NMRD) profiles were recorded on a SpinMaster2000 relaxometer from Stelar Company, Mede, Italy. Measurements of spin–lattice relaxation rates were performed as a function of external magnetic field B_0 strength expressed in Larmor resonance frequency units for protons in the range from 10 kHz to 20 MHz and as a function of temperature. The dependence of the relaxation rate ($R_1 = 1/T_1$) on Larmor frequency is called the dispersion profile and abbreviated as NMRD. The NMRD profiles provide information on motions occurring on time scales from 10^{-4} to 10^{-9} s. Its shape reflects the spectral density function $J(\omega)$, which is a Fourier transform of the correlation function G(t). Time-correlation functions formally describe translational and reorientational motions in physical terms [24]. In the simplest case, the NMR relaxation phenomenon is driven by the stochastic fluctuations of the local magnetic fields and magnetic interactions between the spins, e.g., dipole–dipole interactions. These fluctuations are characterized by a specific time constant called correlation time. The correlation time describes the time scale of losing the information by the molecule (spins) on its initial state. The cause of the fluctuations that determine the relaxation mechanisms are molecular dynamical processes that the molecules undergo [25–27]. In the case of only one dynamical process characterized by a single value of correlation time (referred to as the Bloembergen–Purcell–Pound (BPP) model), the spectral density function can be described as follows [28,29]:

$$J(\omega) = \frac{\tau}{1 + \omega^2\tau^2}$$

where τ is the characteristic time of the fluctuations, the correlation time, and ω is the Larmor frequency. By analyzing the shape of the spectral density function $J(\omega)$, we obtain access to the relaxation mechanism, allowing us to probe the molecular dynamical processes of investigated systems. The $J(\omega)$ function reaches its maximum value when the characteristic time of the local fluctuations equals the reciprocal of the Larmor frequency $\tau_c = \omega^{-1}$.

Depending on the nature of the dynamical processes responsible for the relaxation mechanisms, different theoretical models for analyzing the spectral density functions have been developed (see Section 3.5). For the same molecular motion, the correlation time may differ depending on the temperature, following, e.g., the Arrhenius law, causing the spectral density function to extend toward higher frequencies with decaying amplitude. With this assumption, we can treat the spectral density function as a measure of the probability of spin transitions of a frequency ω induced by the fluctuation of local magnetic fields characterized by the given correlation time τ_c. More macroscopically speaking, this shows that molecular motion described by a time scale of $\tau = \omega^{-1}$ is the most efficient source of relaxation at a given frequency (and thermal conditions) [30].

The advantage of the field-cycling NMR relaxometry method lies in the access to a low-frequency range. Most information on slow dynamical processes characterized with correlation times longer than nanoseconds is reflected in the spectral density at low values of ω, corresponding to magnetic fields below 1T, much below high-resolution NMR spectrometers. Therefore, this method is very useful in studying, e.g., slow collective motions that depend on molecular size and shape, viscosity, temperature, and intermolecular interaction. These features of the NMR FFC technique make it ideal for investigating the molecular dynamics of LC molecules under various mesophases occurring as a function of temperature. The technical details of the FFC NMR relaxometry experiment are reported elsewhere [31,32].

2.2. Thermal Properties and Phase Transition Measurements

The mesophase sequence and phase transition temperature were measured using the differential scanning calorimetry method (DSC) and a Perkin-Elmer DSC4000 instrument (Perkin-Elmer, Waltham, MA, USA). The measurements were performed under a gaseous nitrogen atmosphere in the temperature range from 273 K to 333 K, covering transitions from isotropic to crystal phase. The LC samples were placed in a sealed aluminum crucible and subjected to three consecutive heating and cooling cycles at a scanning rate of 5 °C/min.

2.3. Mesophase Texture Analysis

The mesophase texture images were recorded with an Olympus optical microscope BX53-P (Tokyo, Japan) under a magnification of 200× and crossed polarizers. The samples were placed on a glass microscopic slide and covered with a 0.2 mm cover slip. The sample thickness was 20 μm and the temperature of the sample was stabilized and controlled with a Linkam TSM91 hot stage. The images were taken during the heating cycle.

2.4. Dielectric Spectroscopy Measurements

The dielectric spectroscopy measurements were performed for protonated and deuterated samples with the use of an HP 4192A LF Impedance Analyzer (Hewlett-Packard, Tokyo, Japan) connected with a Mettler FP–82 HT hot stage, at a frequency of 1.1 kHz, oscillation level at 0.1 V, and different BIAS voltages at 5, 10, 15, and 20 Celsius degrees below the I/N phase transition.

3. Results and Discussion

3.1. NMR Spectroscopy Investigations

The selectivity and efficiency of used synthesis routes to prepare partially and fully deuterated 5CB samples were determined by proton NMR spectroscopy studies. The NMR spectrum of the fully protonated 5CB sample was used as a reference. The same volume of all studied LC samples was placed in 5 mm NMR tubes and subjected to measurements at 50 °C, resulting in an isotropic phase. Figure 1 presents recorded spectra for 5CB, 5CB-d_8, and 5CB-d_{19} samples measured under the same conditions without scaling of the spectra. Due to the high sensitivity of the ^{1}H nucleus, even a very small quantity of hydrogen atoms left after deuteration can be easily detected. Based on integral analysis of the ^{1}H spectra, we estimated the degree of conversion for partially and fully deuterated samples and the selectivity of this process. For the 5CB-d_{19} sample, the total deuteration degree was equal to 96%, where the hydrogen atoms in the benzene rings were substituted with deuterons in 93% and the alkyl chain in 97%. In the case of the 5CB-d_8 sample, the total deuteration degree was equal to 47%, where the substitution degree in the aromatic part was equal to 96% and in the side chain to 11%—the result obtained proves that a very high degree of conversion was acquired using the synthesis route. Also, the substitution process selectivity remained at a high level, allowing us to consider the 5CB-d_8 sample as selectively deuterated.

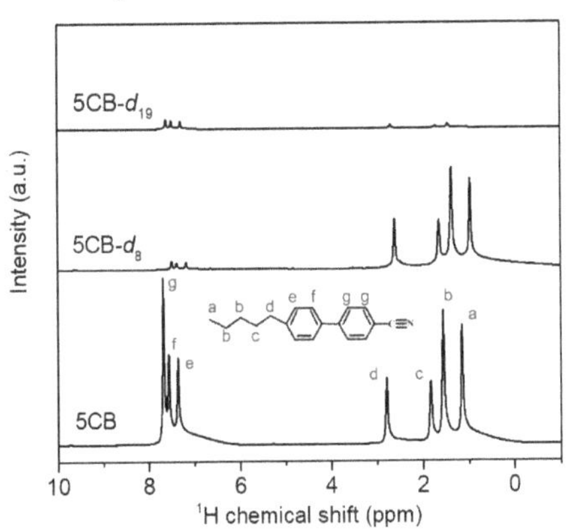

Figure 1. The ^{1}H NMR spectra of 5CB, 5CB-d_8, and 5CB-d_{19} in the isotropic phase.

3.2. Thermal Properties and Phase Transitions—DSC

The thermal properties and phase transition temperatures of 5CB liquid crystal, especially the temperature of the transition from isotropic to nematic phase, are well known in the literature and have been previously determined [33]. In this report, we investigated the thermal properties of 5CB in a broad temperature range and accounted for the transitions from isotropic to crystal phase in order to characterize it as a reference for the same thermal treatment of selectively deuterated 5CB samples. Figure 2 presents the recorded DSC curves for the heating and cooling cycles of the investigated samples. The results clearly show that the degree of deuteration influences the phase transition temperatures and the presence of individual exothermic features. Some similarities can be found for fully protonated and fully deuterated 5CB molecules, whereas for selectively deuterated 5CB-d_8 molecules, the differences in characteristic phase transition temperature are more pronounced; additionally, the partial crystallization observed for 5CB and 5CB-d_{19} during cooling stage is missing. In a fully protonated 5CB molecule (Figure 2a), we can observe endothermic peaks related to crystal-to-nematic (Cr/N) and nematic-to-isotropic (N/I) phase transitions, a glass transition (Tg) feature, and a cold crystallization exothermic peak during heating of the sample. The temperature determined for individual features agrees well with data found in the literature [33]. The most dominant peak observed at 23.7 °C is related to the Cr/N phase transition, followed by the N/I phase transition at 34.8 °C. Upon heating the sample from −120 °C, the transition from glassy to liquid phase at −68.1 °C (Tg—fictive temperature) followed by cold crystallization (exothermic peak at −29.5 °C) was observed. It is worth noticing that the thermal history of the sample affects observed features, e.g., if the sample is cooled after completion of cold crystallization and measured upon heating in the region where the glass transition was previously observed—no glass transition feature is present, proving that the sample was completely crystallized. Therefore, the endothermic peak observed after cold crystallization can be determined as a Cr/N phase transition. During the cooling of the 5CB sample first observed, a sharp exothermic peak at 32.7 °C is related to the transition from an isotropic to a nematic phase (I/N) followed by a broad exothermic peak at −28.5 °C caused by partial crystallization. Subsequent cooling of the sample indicates a glass transition feature related to the transition from the nematic liquid-to-nematic glassy state at −69.1 °C. In the 50 °C to −120 °C temperature range with heating and cooling of the sample at 10 °C/min, no crystallization peak after glass transition in the cooling stage of the 5CB sample was observed in any of the three subsequent heating–cooling cycles.

In the case of the fully deuterated 5CB-d_{19} sample (Figure 2c), the same sequence of endo- and exothermic peaks in heating and cooling cycles with only small differences in the temperatures of particular transitions to the corresponding transitions in the protonated 5CB sample were observed. One additional feature was observed for the 5CB-d_{19} sample during cooling at 10 °C/min, which was detected after the glass transition as a small exothermic peak at −103 °C related to the complete crystallization of the sample. This peak was confirmed in three subsequent cooling–heating cycles in the investigated sample. To explain the same phase transition behavior of the fully protonated 5CB and fully deuterated 5CB-d_{19} samples, we need to consider the fact that complete deuteration of the sample does not introduce any change considering the symmetry and imbalance of the intermolecular interactions between the molecules, which is the case for the partially deuterated 5CB sample.

The most interesting case is observed for a partially deuterated 5CB-d_8 molecule (Figure 2b). In this case, the ^{2}H atoms substituted only ^{1}H atoms in the aromatic core of the molecule, leaving the alkyl chain unchanged. In this case, the mass distribution in the molecule changes, and now, we have a heavy organic core and light alkyl chain. Additionally, the intermolecular interactions are also changed as deuterons have approx. 2.5 times larger size, 3.2 smaller magnetic dipole moment, 2 times larger mass, and same electric charge, but are distributed over a larger volume compared to protons. The first noticeable change with respect to 5CB and 5CB-d_{19} is the greater width of the observed exothermic and endothermic peaks. Next, during the cooling of the 5CB-d_8 sample, no partial crystallization was observed before the glass transition at −67.5 °C. Subsequent cooling leads to a sharp and well-pronounced exothermic peak related to crystallization at −115.2 °C. Upon heating the sample, the analogous sequence of exothermic and endothermic features was observed as in the 5CB and 5CB-d_{19} samples with greater differences in individual phase transition values. Table 2 gathers all the characteristic phase transition temperatures for all the investigated samples measured with the DSC and POM methods.

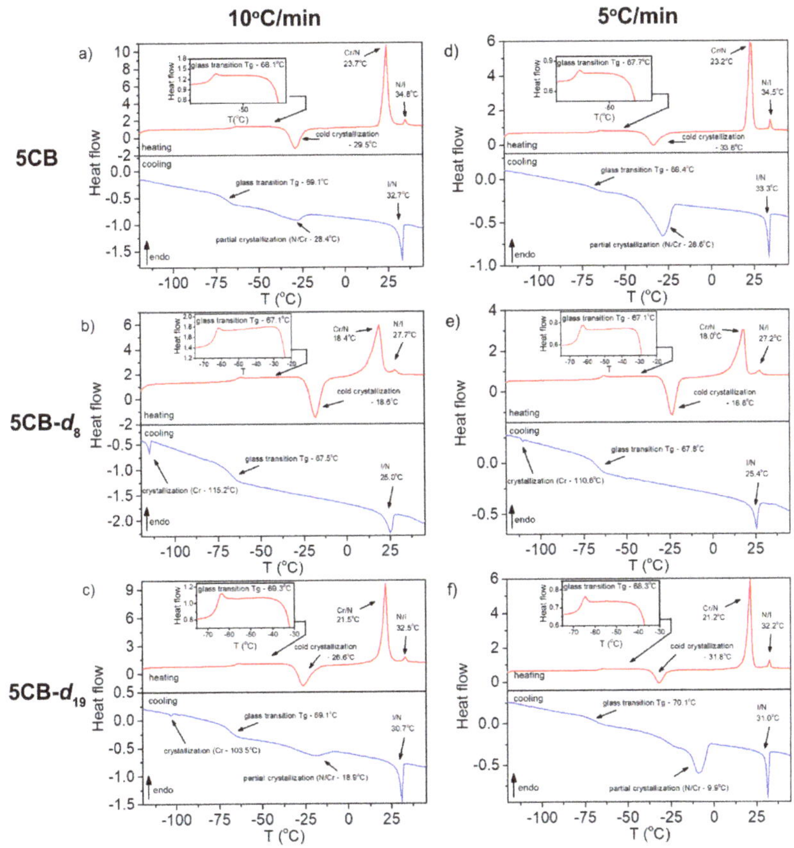

Figure 2. DSC heating and cooling curves for 5CB (**a**), 5CB-d_8 (**b**), and 5CB-d_{19} (**c**) at 10 °C/min and for 5CB (**d**), 5CB-d_8 (**e**), and 5CB-d_{19} (**f**) at 5 °C/min heating rate.

Table 2. The characteristic phase transition temperatures obtained with the DSC and POM method.

Sample	Phase Transition Temperature/°C from DSC								$T_{I/N}$/°C from POM
	Heating				Cooling				Cooling
	Glass Tr	Cold Cr	Cr/N	N/I	I/N	Partial Cr	Glass Tr	Cr	
5CB	−68.1	−29.5	23.7	34.8	32.7	−28.4	−69.1	--	34.2
5CB-d_8	−67.1	−18.6	18.4	27.7	25.0	--	−67.5	−115.2	27.6–25.8
5CB-d_{19}	−69.3	−26.6	21.5	32.5	30.7	−18.9	−69.1	−103.5	32.0–31.7

Glass Tr—glass transition; Cold Cr—cold crystallization, Cr/N—crystal-to-nematic phase transition; N/I—nematic-to-isotropic phase transition; I/N—isotropic-to-nematic phase transition; Partial Cr—partial crystallization; Cr—crystallization.

All samples were measured for two temperature change rates, 10 °C/min and 5 °C/min. With the decrease in the heating/cooling rate, the temperatures for individual phase transitions showed only small differences except for cold crystallization in fully protonated and fully deuterated samples and partial crystallization in fully deuterated samples, where the phase transition temperature changes were more pronounced. This interesting effect is worth future investigation with more heating/cooling change rates and isothermal measurements.

In summary, when the 5CB is cooled at 10 °C/min, the sample is in the isotropic phase at temperatures above 33 °C, in the nematic phase in the temperature range from −25 °C to 33 °C, and in supercooled nematic liquid with a small fraction of crystallites below −25 °C. Upon heating, the 5CB behaves as supercooled nematic liquid up to −30 °C, in the temperature range from −30 °C to 24 °C it is in crystalline form, in the nematic phase in the range from 24 °C to 35 °C, and finally in isotropic phase above 35 °C.

For the 5CB-d_8 cooled at 10 °C/min, the sample exists in an isotropic phase above 25 °C, in the nematic/supercooled nematic phase for temperatures between −60 °C and 25 °C, and in glassy nematic liquid in the temperature range from −115 °C to −60 °C, with crystallization at −115 °C. Upon heating, the sample behaves as supercooled nematic liquid up to −19 °C and as crystalline in the temperature range between −19 °C and 19 °C. It exists in the nematic phase between 19 °C and 28 °C and in the isotropic phase above 28 °C.

The 5CB-d_{19} during cooling at 10 °C/min exists in the isotropic phase above 31 °C, in the nematic phase in the temperature range from −19 °C to 31 °C, and as the supercooled nematic liquid with a fraction of crystallites below −19 °C, with crystallization at −103 °C. During heating, the 5CB-d_{19} behaves as supercooled nematic liquid up to −27°C, as crystalline between −27 °C and 21 °C, as nematic in the temperature range from 21 °C to 32.5 °C, and as an isotopic phase above 33 °C.

3.3. Mesophase Texture Analysis

To investigate the influence of deuteration on the intermolecular interactions and self-organization of the 5CB, 5CB-d_8, and 5CB-d_{19} molecules, we recorded and compared the LC textures in the temperature range of the nematic phase existence. Substantial differences between protonated and all deuterated samples were observed. All recorded textures are presented in Figure 3. In the case of the protonated 5CB sample on reaching the temperature of nematic phase existence, the texture's uniform color was observed, confirming the homogeneous alignment of the LC molecules throughout the cell. Further, decreasing the temperature reduced the thermal vibrations and led to an increase in the ordering parameter, which is reflected by the change in the texture color. In the case of the 5CB-d_8 sample, the domain texture was observed upon reaching the transition from isotropic to the nematic phase (I/N). At proximity to the I/N phase transition, small and separated regions of the nematic phase were homogeneously spread in the continuous isotropic phase. Further decrease in the temperature caused the growth of the nematic domains, which gradually started to be in contact with its boundaries. At this stage, the volume of the isotropic phase was decreased but remained continuous. Further decrease in temperature led to merging of the individual nematic domains and vanishing of the domain walls. The trace of vanishing domain walls is seen as a gradual change in the texture's color in merged domains. For sufficiently large domains, the same gradient of the texture color is observed close to domain walls. This effect indicates that the alignment of the 5CB-d_8 molecules is different at domain walls and in the center of the domain. Further decreases in temperature lead to the merging of all nematic domains, making the isotropic phase become single and separated domains that finally disappear. After the disappearance of the domains of the isotropic phase, the color of the texture of the nematic phase becomes uniform throughout the whole cell. This effect indicates that the alignment of the 5CB-d_8 molecules becomes uniform in the entire volume. In the case of fully deuterated 5CB-d_{19}, the domain texture of nematic regions occurred at the onset of the I/N phase transition. Further decrease in temperature leads to the growth of the nematic domains, which eventually merge. During the development of the nematic phase domains, the isotropic regions follow the behavior observed in 5CB-d_8. However, in the case of 5CB-d_{19} molecules, no gradient of the texture color in nematic domains was observed even after merging. Therefore, no trace of the vanishing domain walls was observed. This indicates that alignment in the nematic domains of 5CB-d_{19} molecules remains uniform and homogenous, close and far from the domain walls. A change in the texture color for

5CB-d_{19} was observed upon cooling after the domains of the isotropic phase had vanished, meaning that 5CB-d_{19} follows the order parameter behavior as a function of temperature as in other studied 5CB samples. It is worth noticing that the whole process of occurrence and growth of the nematic phase was completed in a much narrower temperature range than for 5CB-d_8.

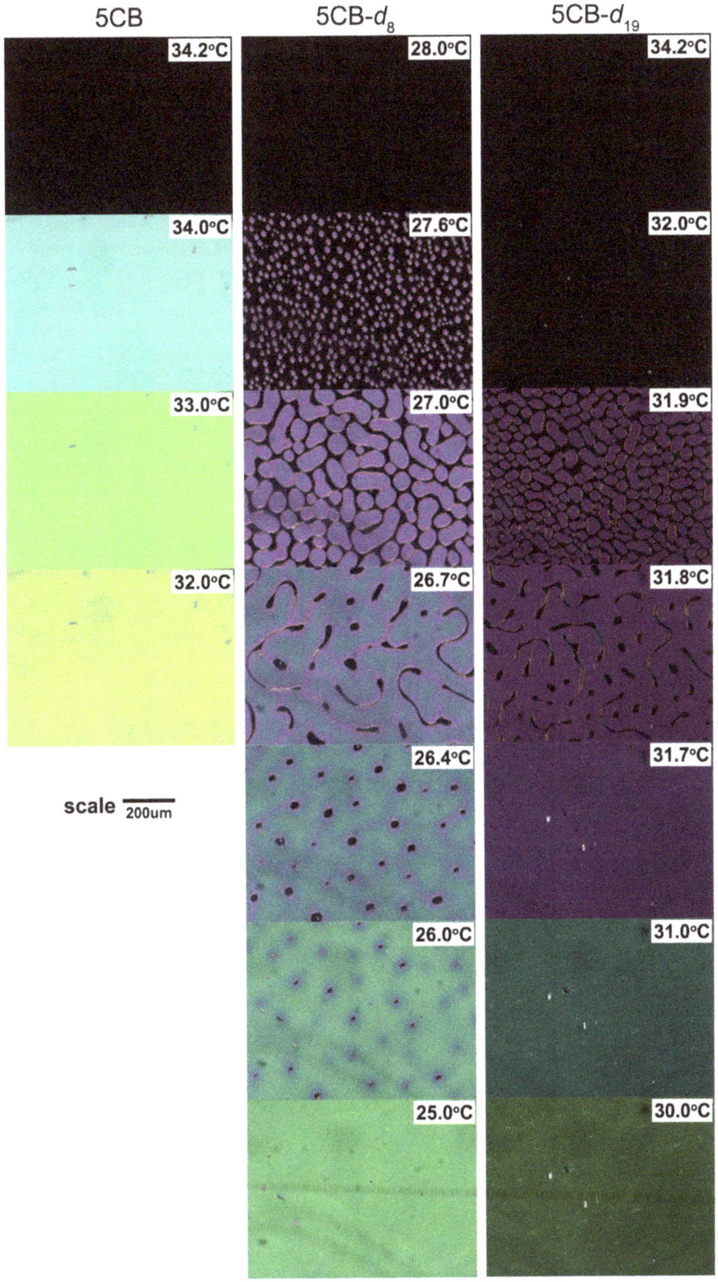

Figure 3. Polarization microscopy images of 5CB, 5CB-d_8, and 5CB-d_{19} liquid crystal.

3.4. Dielectric Spectroscopy Measurements

In dielectric liquid crystals, by applying an external electric field, it is possible to reorient the LC molecules according to the direction of the applied field. In nematic liquid crystals, the direction of the optical axis is described with the director $\boldsymbol{n}$, representing the average unit vector along the molecular axis. By changing the director's orientation, the optical properties are also changed. In the absence of the external electric field, the orientation of n is determined by the external interactions and anchoring effect. Applying the electric field, which is initially perpendicular to the director, causes the deformation of a homogenous layer of LC molecules above a specific strength of the electric field, called the threshold voltage. For sufficiently high fields, it is possible to completely reorient the director from a perpendicular to parallel orientation. This effect is possible due to the anisotropy of the dielectric permittivity $\Delta\varepsilon = \varepsilon_{\parallel} - \varepsilon_{\perp}$, where $\varepsilon_{\perp}$ and $\varepsilon_{\parallel}$ is the electric permittivity measured along the directions perpendicular and parallel to n, respectively. With positive dielectric anisotropy ($\Delta\varepsilon > 0$), the director is forced to align along the electric field. To investigate the influence of selective deuteration on the dielectric properties of studied 5CB derivatives and interactions between the studied molecules, we analyzed the dependence of the electric permittivity ε' on the BIAS electric field. Figure 4 shows recorded profiles for the 5CB, 5CB-d_8, and 5CB-d_{19} samples at different temperatures within the nematic phase. Based on the results, we can conclude that partial deuteration influences the dielectric permittivity and threshold voltage of investigated samples the most. The calculated parameters are gathered in Table 3. The possible source of this effect is related to changes in charge distribution over the 5CB-d_8 molecules and D/H intermolecular interactions, which are stronger than H/H interactions. Moreover, the DSC investigations revealed additional phase transitions within the nematic phase, which can induce more complex interaction with the external electric field due to different molecule alignments. Figure 5 shows the direct comparison of the dielectric permittivity ε' vs. BIAS electric field at a given reduced temperature $T_{I/N}$ for all the samples, showing similarities between fully protonated and deuterated 5CB molecules, which substantially differ from partially deuterated 5CB molecules. In both cases, we attribute this behavior to charge distribution and intermolecular interaction symmetry. To deliver more quantitative results for these interactions, high-resolution NMR spectroscopy measurements at high fields for ^{1}H and ^{2}H atoms are planned.

Table 3. Values of $\Delta\varepsilon$ and U_{th} for 5CB, 5CB-d_8, and 5CB-d_{19} at different temperatures in the nematic phase.

Sample	**$\Delta\varepsilon$ (1.1 kHz, $T_{I/N}$-5 °C)**	**U_{th} (1.1 kHz, $T_{I/N}$-5 °C)**
5CB	10.5	4.4
5CB-d_8	9.1	6.6
5CB-d_{19}	10.1	3.9
	$\Delta\varepsilon$ (1.1 kHz, $T_{I/N}$-10 °C)	**U_{th} (1.1 kHz, $T_{I/N}$-10 °C)**
5CB	11.9	4.4
5CB-d_8	10.7	6.5
5CB-d_{19}	11.4	4
	$\Delta\varepsilon$ (1.1 kHz, $T_{I/N}$-15 °C)	**U_{th} (1.1 kHz, $T_{I/N}$-15 °C)**
5CB	12.8	4.4
5CB-d_8	11.7	6.9
5CB-d_{19}	12.2	4.3
	$\Delta\varepsilon$ (1.1 kHz, $T_{I/N}$-20 °C)	**U_{th} (1.1 kHz, $T_{I/N}$-20 °C)**
5CB	13.3	4.4
5CB-d_8	12.5	6.9
5CB-d_{19}	12.8	5

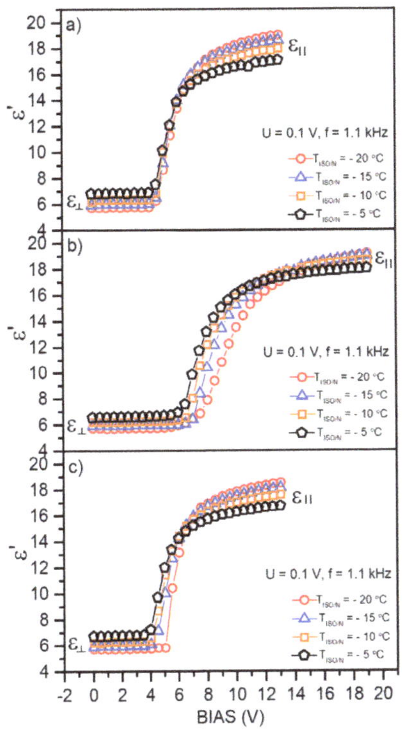

Figure 4. Electrical permittivity as a function of the voltage of the BIAS external field for the nematic phase in the studied materials: (**a**) 5CB, (**b**) 5CB-d_8, and (**c**) 5CB-d_{19}.

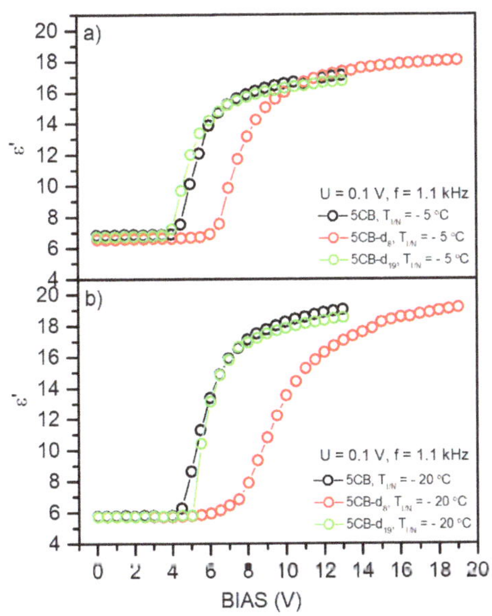

Figure 5. Electrical permittivity ε dependence upon the BIAS electric field for the 5CB, 5CB-d_8, and 5CB-d_{19} materials at reduced temperature $T_{I/N}$ −5 °C (**a**) and −20 °C (**b**).

3.5. NMR Relaxometry Study

The spin–lattice relaxation studies of protonated and deuterated samples were performed at different temperatures in isotropic and nematic phases with the use of the fast field cycling nuclear magnetic resonance relaxometry method in the 10 kHz–20 MHz frequency range and at a high field (500 MHz). For the 5CB-d_{19} sample, only the isotropic phase was investigated because the dramatic shortening of spin–spin relaxation time prevents us from recording measurable signals in the nematic phase. Figure 6 shows recorded NMRD profiles for investigated samples in isotropic and nematic phases at different temperatures. In the case of the 5CB and 5CB-d_8 samples, the measurements were performed on the ^{1}H nucleus in the isotropic and nematic phases. The ^{2}H NMRD profiles were recorded for the 5CB-d_8 and 5CB-d_{19} samples in the isotropic phase. The too-short T_2 relaxation times for ^{2}H upon crossing to nematic phases did not allow us to use this nucleus to record ^{2}H NMRD profiles below the I/N phase transition temperature.

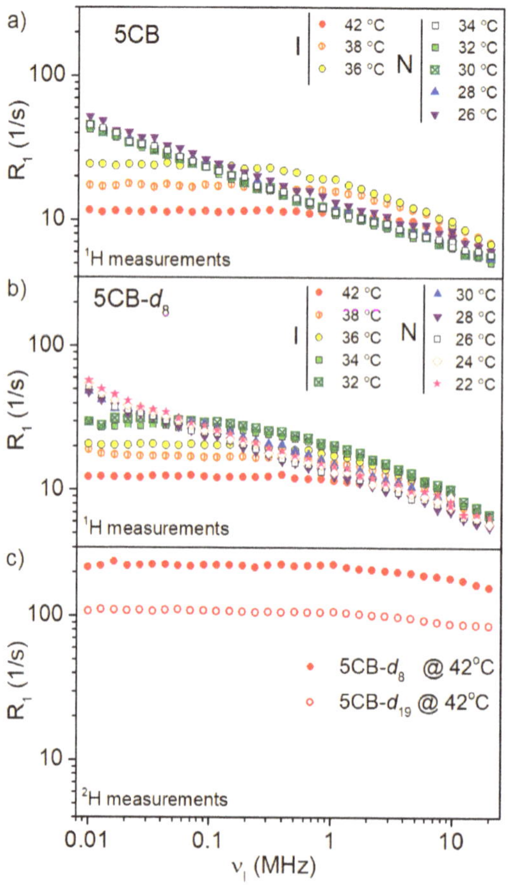

Figure 6. ^{1}H relaxation dispersion profiles for (**a**) 5CB and (**b**) 5CB as a function of temperature together with ^{2}H relaxation dispersion profiles for 5CB-d_8 and 5CB-d_{19} in the isotropic phase (**c**).

In all studied cases, we observed monoexponential dependence of the magnetization decay or recovery, which proves that the external magnetic field was not large enough to influence the mesogenic molecules' alignment. A single T_1 relaxation time was fitted to the experimental data at all studied frequencies. The NMRD profiles were analyzed according to theoretical approaches used for the description of the relaxation mechanisms, as follows:

- Local rotations/reorientations (R) of LC molecule around the molecular axis; the Woessner model with appropriate modifications was used to describe the dynamics in the isotropic and nematic phases.
- Translational self-diffusion (SD) is a type of motion that can be affected by local structure. In the isotropic phase, we used the Torrey model for the isotropic self-diffusion constant. In the nematic phase, the model was extended using the Zumer and Vilfan approach to account for anisotropy of the D constant parallel and perpendicular.
- Order director fluctuations (ODF) describe the orientational fluctuations of nematic director *n* with respect to its time-average orientation. This model reflects the collective reorientational motions of the mesogenic molecules.
- The local order fluctuations (OF) are a contribution that can occur in the isotropic phase just above the N/I temperature, showing square-root power law dependence. Both ODF and OF are collective-type motions that can give similar fits but with different model parameter values.

The general equation used to analyze the NMRD profiles of studied LC systems considers the sum of contributions to relaxation rates R_1 accounting for particular mesophase structures and types of motions. The contributions are statistically independent from each other or have distinct time scales.

$$\left(\frac{1}{T_1}\right)_{Iso} = \left(\frac{1}{T_1}\right)_R + \left(\frac{1}{T_1}\right)_{SDi} + \left(\frac{1}{T_1}\right)_{ODF|OF}. \tag{1}$$

We used the following models for the isotropic phase to describe the molecular dynamics. For the spin–lattice relaxation contribution by local rotations/reorientations (R) around the long (z) and short (x) molecular axes, the relaxation rate is given by the Nordio model [34]:

$$\left(\frac{1}{T_1}\right)_R = \frac{3}{4}K_D\left[J_R^{(1)}(\omega_L) + J_R^{(2)}(2\omega_L)\right], \tag{2}$$

where $\omega_L = 2\pi\nu_L$ is the Larmor frequency, and the spectral density function, $J_R{}^{(k)}$ (for k = 1, 2), is written as:

$$J_R^{(k)} = \frac{4}{3}(k)^2\sum_{m=0}^{2}\frac{\overline{\left|d_{m,0}^2(\alpha_{ij})\right|}^2}{r_{ij}^6} \times c(k,\, m)\frac{\left(\tau_{k,\,m}^2\right)^{-1}}{k^2\omega_0^2 + \left(\tau_{k,\,m}^2\right)^{-2}} \tag{3}$$

where $\tau_{k,m}{}^2$ depends on correlation times for rotations around a short (τ_x) and long molecular axis (τ_z):

$$\left(\tau_{k,\,m}^2\right)^{-1} = \tau_z\left[\frac{1}{\beta_{k,m}^2} + \left(\frac{\tau_x}{\tau_z} - 1\right)m^2\right]. \tag{4}$$

In Equations (3) and (4), $c(k,\,m)$ and $\beta_{k,m}{}^2$ are numerical functions, and $d_{m,0}^2(\alpha_{ij})$ are the reduced Winger matrices [35].

Torrey's model, associated with the translational self-diffusion for isotropic-like systems, can be written in the form [36,37]

$$\left(\frac{1}{T_1}\right)_{SD_i} = \frac{9}{8}\gamma^4\hbar^2\left(\frac{\mu_0}{4\pi}\right)^2\frac{n\tau_D}{d^3} \times [\mathcal{T}(\alpha,\omega\tau_D) + 4\mathcal{T}(\alpha,2\omega\tau_D)], \tag{5}$$

where d is the closest distance between the molecules, τ_D is the average time between molecular translational jumps n–^{1}H spin density, $\alpha = \langle a^2\rangle/12d^2$, and $\langle a^2\rangle = 6\tau_D D_i$ is the mean square root of the molecular jump distance, D_i is the diffusion constant, and $\mathcal{T}(\alpha, x)$ are analytical functions [36,37].

The OF relaxation contribution can be described by the following expression [37]

$$\left(\frac{1}{T_1}\right)_{OF} = \frac{A_{OF}}{\omega_L^{1/2}} \int_0^{\nu_{cM}^{OPF}/\nu_L} \frac{\sqrt{x}}{1+\left(x+\nu_0^{OF}/\nu_L\right)^2} dx, \tag{6}$$

where A_{OF}, ν_{cM}^{OF}, and ν_0^{OF} depend on viscoelastic parameters in the isotropic phase and denote the strength of the OF process at the high and low cutoff frequency.

In the isotropic phase for 5CB and 5CB-d_8 liquid crystal, the ^{1}H and ^{2}H NMRD profiles show similar shapes. In the low-frequency region, a plateau was observed. The R_1 values at this plateau vary with temperature from 11.5 to 24 s^{-1} in 5CB and 12 to 30 s^{-1} in 5CB-d_8, by cooling from 42 to 36 °C and 42 to 32 °C, respectively.

To describe the intermolecular interactions and motions in the nematic phase, we used the same general approach given in Equation (1) but with appropriate modifications for the given molecular processes. In the case of molecular rotations/reorientations along the molecular axis, we used the same Nordio model as in the case of the isotropic phase. For the description of translational self-diffusion (SD) processes, we used the Torrey model expanded by Žumer and Vilfan, who took into account the anisotropy of the system. As a result, the following expression for the spin–lattice relaxation contribution related to self-diffusion was considered [38]:

$$\left(\frac{1}{T_1}\right)_{SD_a} = \frac{9}{8}\gamma^4\hbar^2\frac{n\tau_D}{d^3} \times Q\left(\omega_L\tau_D, \frac{\langle a_\perp^2\rangle}{d^2}, \frac{l}{d}, \frac{D_\parallel}{D_\perp}\right), \tag{7}$$

where $\langle a_\perp^2\rangle = 4\tau_{D_\perp}D_\perp$ is the width of the molecule, l is the layer thickness, n is the spin density, $\tau_{D_\perp}$ is the mean square of molecular jump time, and $D_\perp$ and $D_\parallel$ are diffusion constants of molecules in two perpendicular directions to the nematic director; Q is a dimensionless function calculated numerically. In the nematic phase, we also used the model for order director fluctuations (ODF) instead of local order fluctuations as in the I phase.

$$\left(\frac{1}{T_1}\right)_{ODF} = \frac{A_{ODF}}{\omega^{1/2}}\left[f\left(\frac{\omega_{ODFmax}}{\omega}\right) - f\left(\frac{\omega_{ODFmin}}{\omega}\right)\right] \tag{8}$$

where $f(x)$ is the cutoff function and the prefactor:

$$A_{ODF} = (9/8)(\mu_0/4\pi)^2\gamma^4\hbar^2k_BTS^2\eta^{1/2}/\left(\sqrt{2}\pi K^{3/2}a_{eff}^6\right) \tag{9}$$

where a_{eff}^6 depends on the interproton distances and interproton vector angles with respect to the long molecular axis. η and K are the viscosity and elastic constant of nematic liquid crystal.

The NMRD profiles measured in the nematic phase change significantly as to the isotropic phase, showing strong dispersion dependence in the low-frequency range (Figure 5). Figure 7 shows the best fits of the mentioned models to experimental data for 5CB at 42 °C and 5CB-d_8 at 34 °C in the isotropic phase and at 30 °C and 22 °C in the nematic phase, respectively. The experimental data also included R_1 values measured at 500 MHz, which were of great importance in the analysis of the results because they allowed for a more faithful reproduction of the shape of the relaxation profiles. The fitting parameters are gathered in Tables 4 and 5.

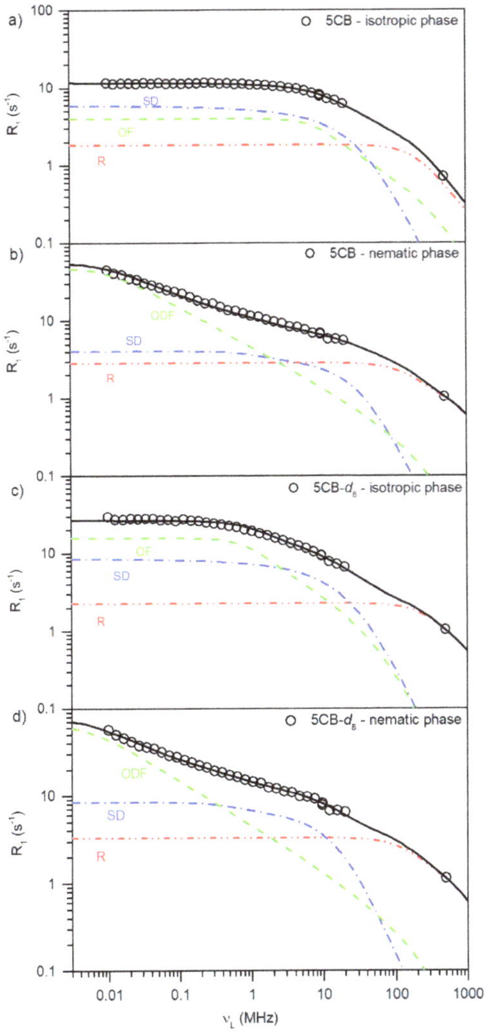

Figure 7. ^{1}H and ^{2}H experimental NMRD profiles R_1 for 5CB in isotropic phase (**a**) and nematic phase (**b**) and for 5CB-d_8 in isotropic phase (**c**) and nematic phase (**d**), together with the best fits obtained with the model given by Equation (1).

Table 4. The isotropic phase, where the order parameter $S_0 = 0$, low cutoff frequency $\omega_{cmin}/2\pi = 0$ Hz, the geometrical factors $A_0 = 6 \times 10^{57}$ m^{-6}, $A_1 = 3 \times 10^{57}$ m^{-6}, $A_2 = 9 \times 10^{57}$ m^{-6}, the distance of the closest approach $d = 5 \times 10^{-10}$ m, and $n = 4.6 \times 10^{22}$.

Fitting Parameter	5CB, 42 °C	5CB-d_8, 34 °C	1/T_1 Model
τ_z/s	3.30×10^{-10}	4.67×10^{-9}	R
τ_x/s	5.02×10^{-8}	5.48×10^{-8}	R
D/m^{2}s^{-1}	1.59×10^{-11}	1.08×10^{-11}	SDi
A_{OF}/s$^{-3/2}$	1.04×10^4	1.30×10^4	OF
ν_0^{OF}/Hz	4.59×10^5	1.59×10^4	OF
ν_{cM}^{OF}/Hz	9.00×10^8	1.00×10^8	OF

Table 5. The nematic phase, where the order parameter $S_0 = 5$ and the geometrical factors A_0, A_1, A_2, d, and n take the same values as in the isotropic phase.

Fitting Parameter	5CB, 30 °C	5CB-d_8, 22 °C	1/T_1 Model
τ_Z/s	4.05×10^{-9}	3.14×10^{-9}	R
τ_X/s	1.84×10^{-8}	1.68×10^{-8}	R
τ_D/s	3.09×10^{-9}	6.43×10^{-9}	SDa
$A_{ODF}/s^{-3/2}$	4.78×10^{3}	4.69×10^{3}	ODF
ν_{cm}^{OF}/Hz	8.00×10^{3}	4.31×10^{3}	ODF
ν_{cM}^{OF}/Hz	5.00×10^{8}	3.88×10^{8}	ODF

The results show that the fully protonated 5CB liquid crystal is characterized by smaller order fluctuations and a faster self-diffusion coefficient compared to 5CB-d_8. These results stay in agreement with conclusions obtained from other methods for studied systems, reflecting the nonuniform alignment of 5CB-d_8 molecules close to $T_{I/N}$ and additional phase transitions in the partially deuterated sample, smaller mass of the molecules, and weaker intermolecular interactions.

Figure 8 shows the dependence of the relaxation rate R_1 in different phases for selected Larmor frequencies. The dashed gray line indicates the I/N phase transition temperature determined from DSC measurement upon cooling. Within the isotropic phase, the R_1 values increase with a decrease in the temperature at all Larmor frequencies for the 5CB and 5CB-d_8 samples. At the temperature of the I/N phase transition and further, a change in the R_1 dependence is observed. For lower frequencies ($\leq$100 kHz), an abrupt slope change is observed at the I/N transition, followed by almost constant R_1 throughout the N phase for 5CB, whereas an increase in R_1 value is observed for 5CB-d_8. Moreover, in the case of the 5CB-d_8 sample below the I/N phase transition temperature, the NMRD profile recorded at 100 kHz frequency shows a kind of curvature that is not seen for 5CB. This can originate from the pretransition effect when the system prepares to create the I/N domain structure, evidenced by the POM texture images for the 5CB-d_8 sample before reaching the uniform nematic phase. It is worth noticing that DSC measurements did not detect this state as it is not related to the change in the heat flow in the sample (or the change is too small to be detected in the applied experimental conditions) but with the change of molecular alignment and local self-organization.

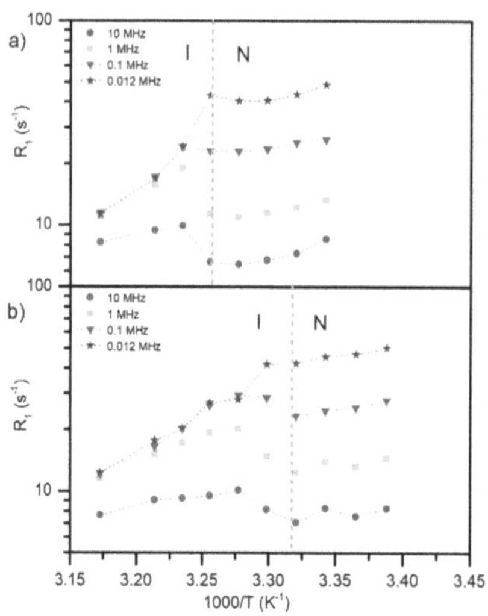

Figure 8. Temperature dependence of R_1 at chosen ^{1}H Larmor frequencies. The vertical dashed lines correspond to the phase transition determined by DSC on cooling [39] for 5CB (**a**) and 5CB-d_8 (**b**).

4. Conclusions

In this work, we investigated the influence of selective deuteration of 5CB liquid crystals on molecular dynamics, phase transitions, textures, and dielectric properties. Deuteration plays an important role in chemical modifications of liquid crystal materials utilized in near-infrared applications to reduce common disadvantages, like parasitic absorption of LCs. Our studies have shown that a fully deuterated 5CB sample exhibits the same phase transition sequence as fully protonated material with almost the same range of nematic phase occurrence shifted only by 3 °C lower compared to 5CB. The protonated 5CB molecules exhibit a sharp transition from isotropic to nematic phase with uniform alignment of the molecules. In the 5CB-d_{19} sample, we detected a nematic/isotropic domain structure at the I/N transition, which transforms into a homogenous nematic phase with uniform molecule alignment within 0.3 °C. Moreover, the dielectric properties and the dependence of the dielectric permittivity in the function of the applied electric field and the threshold field are very similar to those of 5CB. Significant differences in properties were detected in partially deuterated 5CB molecules (5CB-d_8). The thermal analysis has shown a similar sequence of thermal events during the heating stage but with big differences in individual temperatures of the phase transitions. During the cooling of the 5CB-*d8* sample, the differences in phase transition temperature were also significant. Additionally, the sample showed no partial crystallization in the nematic phase before the glass transition temperature. The characteristic I/N phase transition temperature is shifted by 10 °C lower as protonated 5CB molecules and 6 °C lower as fully deuterated 5CB-d_{19} molecules. The texture analysis showed a domain structure that spans over 2 °C with no uniform alignment of the molecules. Moreover, the dependence of the dielectric permittivity in the function of the external electric field differs significantly for both fully protonated and deuterated 5CB samples. The FFC NMR measurements performed indicate the high sensitivity of this method, especially at low Larmor frequencies, for the detection of pretransitional effects and its influence on the molecular dynamics in the investigated samples, not seen by other commonly used experimental techniques for the study of liquid crystals. In conclusion, we state that deuteration can be a suitable method for optimizing the LC properties for utilization in NIR applications, but the process has to be performed thoroughly. Partial deuteration introduces changes in the phase transition sequence, alters dielectric properties and molecular dynamics, and causes the appearance of the domain structure in the nematic phase close to the I/N transition. The possible origin of such behavior is in the change in the symmetry of charge distribution, mass, and intermolecular interaction in partially deuterated systems.

Author Contributions: J.T.-G.: validation, writing—review and editing, and supervision; M.K.: investigation, formal analysis, validation, and writing—review and editing; P.H.: resources, methodology, and investigation; J.H.: resources, methodology, and investigation; M.B.: conceptualization, formal analysis, validation, visualization, writing—original draft, writing—review and editing, and supervision. All authors have read and agreed to the published version of the manuscript.

Funding: This work was partially supported by the National Center for Research and Development, grant DOB-1P/01/03/2016.

Data Availability Statement: The raw data supporting the conclusions of this article will be made available by the authors on request.

Conflicts of Interest: The authors declare no conflicts of interest.

References

1. Yang, K.L.; Cadwell, K.; Abbott, N.L. Use of self-assembled monolayers, metal ions and smectic liquid crystals to detect organophosphonates. *Sens. Actuators B Chem.* **2005**, *104*, 50–56. [CrossRef]
2. Prakash, J.; Parveen, A.; Mishra, Y.K.; Kaushik, A. Nanotechnology-assisted liquid crystals-based biosensors: Towards fundamental to advanced applications. *Biosens. Bioelectron.* **2020**, *168*, 112562. [CrossRef]
3. White, T.J.; Cazzell, S.A.; Freer, A.S.; Yang, D.-K.; Sukhomlinova, L.; Su, L.; Kosa, T.; Taheri, B.; Bunning, T.J. Widely tunable, photoinvertible cholesteric liquid crystals. *Adv. Mater.* **2011**, *23*, 1389–1392. [CrossRef]

4. Cervini, R.; Simon, G.P.; Ginic-Markovic, M.; Matisons, J.G.; Huynh, C.; Hawkins, S. Aligned silane-treated MWCNT/liquid crystal polymer films. *Nanotechnology* **2008**, *19*, 175602. [CrossRef] [PubMed]
5. Keum, C.; Becker, D.; Archer, E.; Bock, H.; Kitzerow, H.; Gather, M.C.; Murawski, C. Organic Light-Emitting Diodes Based on a Columnar Liquid-Crystalline Perylene Emitter. *Adv. Opt. Mater.* **2020**, *8*, 2000414. [CrossRef]
6. Hou, J.G.; Zhao, W.; Zhou, G.; Liu, D.; Broer, D.J.; Feringa, B.L.; Chen, J. Phototriggered Complex Motion by Programmable Construction of Light-Driven Molecular Motors in Liquid Crystal Networks. *JACS* **2022**, *144*, 6851–6860. [CrossRef]
7. Cherian, S.; Donley, C.; Mathine, D.; LaRussa, L.; Xia, W.; Armstrong, N. Effects of field dependent mobility and contact barriers on liquid crystalline phthalocyanine organic transistors. *J. Appl. Phys.* **2004**, *96*, 5638–5643. [CrossRef]
8. Coles, H.; Morris, S. Liquid-crystal lasers. *Nat. Photon.* **2010**, *4*, 676–685. [CrossRef]
9. Humar, M.; Musevic, I. 3D microlasers from self-assembled cholesteric liquid-crystal microdroplets. *Opt. Express* **2010**, *18*, 26995–27003. [CrossRef]
10. Yokoyama, S.; Mashiko, S.; Kikuchi, H.; Uchida, K.; Nakamura, T. Laser emission from a polymer-stabilized liquid-crystalline blue phase. *Adv. Mater.* **2006**, *18*, 48–51. [CrossRef]
11. Xu, J.; Yang, R.; Fan, Y.; Fu, Q.; Zhang, F. A Review of Tunable Electromagnetic Metamaterials with Anisotropic Liquid Crystals. *Front. Phys.* **2021**, *9*, 633104. [CrossRef]
12. Reinitzer, F. Beiträge zur Kenntniss des Cholesterins. *Monatshefte Für Chem.* **1888**, *9*, 421–441. [CrossRef]
13. Wu, S.T. Infrared Properties of Nematic Liquid-Crystals—An Overview. *Opt. Eng.* **1987**, *26*, 120–128. [CrossRef]
14. Wu, S.T.; Cox, R.J. Potential infrared liquid crystals. *Liq. Cryst.* **1989**, *5*, 1415–1424. [CrossRef]
15. Pytlarczyk, M.; Gaczoł, K.; Harmata, P.; Dziaduszek, J.; Herman, J. Nematic stability of 2,2′,4-trifluoro-4″-alkyl-[1,1′:4′,1″]terphenyls and its deuterated isotopologue. *J. Mol. Liq.* **2021**, *335*, 116162. [CrossRef]
16. Dobson, C.A.; Morris, S.M.; Qasim, M.M.; Gardiner, D.J.; Wilkinson, T.D.; Coles, H.J. Wavelength-tuneable liquid crystal lasers from the visible to the near-infrared. *Proc. SPIE* **2011**, *8114*, 81140T-1. [CrossRef]
17. Cheng, Z.; Wang, T.; Li, X.; Zhang, Y.; Yu, H. NIR–Vis–UV light-responsive actuator films of polymer-dispersed liquid crystal/graphene oxide nanocomposites. *ACS Appl. Mater. Interfaces* **2015**, *7*, 27494–27501. [CrossRef] [PubMed]
18. Wu, S.T.; Wang, Q.-H.; Kempe, M.D.; Kornfield, J.A. Perdeuterated cyanobiphenyl liquid crystals for infrared applications. *J. Appl. Phys.* **2002**, *92*, 7146–7148. [CrossRef]
19. Gray, G.W.; Mosley, A. The Synthesis of Deuterated 4-n-Alkyl-40-Cyanobiphenyls. *Mol. Cryst. Liq. Cryst.* **1978**, *48*, 233–242. [CrossRef]
20. Kula, P.; Herman, J.; Harmata, P.; Czerwiński, M. NIR, and MWIR transparent liquid crystals. In Proceedings of the 2014 39th International Conference on Infrared, Millimeter, and Terahertz Waves (IRMMW-THz), Tucson, AZ, USA, 14–19 September 2014; pp. 1–2. [CrossRef]
21. Kula, P.; Bennis, N.; Marć, P.; Harmata, P.; Gacioch, K.; Morawiak, P.; Jaroszewicz, L.R. Perdeuterated liquid crystals for near-infrared applications. *Opt. Mater.* **2016**, *60*, 209–213. [CrossRef]
22. Zak, M.; Herman, J.; Czerwiński, M.; Arakawa, Y.; Tsuji, H.; Kula, P. Deuterated Liquid Crystals—Design and synthesis of deuterium-labeled 4,4″-dialkyl-2′,3′-difluoro-[1,1′:4′,1″]terphenyls using batch and continuous flow systems. *Liq. Cryst.* **2022**, *50*, 622–635. [CrossRef]
23. Pytlarczyk, M.; Herman, J.; Arakawa, Y.; Tsuji, H.; Kula, P. Deuterated Liquid Crystals—Practical synthesis of deuterium-labeled 4-alkyl-4″-isothiocyanato-[1,1′:4′,1″]terphenyls. *J. Mol. Liq.* **2022**, *345*, 117847. [CrossRef]
24. Williams, G. Time-correlation functions and molecular motion. *Chem. Soc. Rev.* **1978**, *7*, 89–131. [CrossRef]
25. Levit, M.H. *Spin Dynamics: Basics of Nuclear Magnetic Resonance*, 2nd ed.; John Wiley & Sons Ltd.: West Sussex, UK, 2008.
26. Kruk, D. *Understanding Spin Dynamics*, 1st ed.; Pan Stanford Publishing Pte Ltd.: Olsztyn, Poland, 2016.
27. Khashami, F. *Fundamentals of NMR and MRI: From Quantum Principles to Medical Applications*, 1st ed.; Springer Nature: Cham, Switzerland, 2024.
28. Anoardo, E.; Galli, G.; Ferrante, G. Ferrante Fast-Field-Cycling NMR: Applications and Instrumentations. *Appl. Magn. Reson.* **2001**, *20*, 365–404. [CrossRef]
29. Parigi, G.; Ravera, E.; Fragai, M.; Luchinat, C. Unveiling protein dynamics in solution with field-cycling NMR relaxometry. *Prog. Nucl. Magn. Reson. Spectrosc.* **2021**, *124–125*, 85–98. [CrossRef]
30. Korb, J.-P. Multiscale nuclear magnetic relaxation dispersion of complex liquids in bulk and confinement. *Prog. Nucl. Magn. Reson. Spectrosc.* **2018**, *104*, 12–55. [CrossRef]
31. Noack, F. NMR field-cycling spectroscopy: Principles and applications. *Prog. Nucl. Magn. Reson. Spectrosc.* **1986**, *18*, 171–276. [CrossRef]
32. Kimmich, R.; Anoardo, E. Field-cycling NMR relaxometry. *Prog. Nucl. Magn. Reson. Spectrosc.* **2004**, *44*, 257–320. [CrossRef]
33. Kripotou, S.; Georgopoulos, D.; Kyritsis, A.; Pissis, P. Phase Transitions and Molecular Mobility in 5CB and CE8 Studied by Dielectric Techniques. *Mol. Cryst. Liq. Cryst.* **2015**, *623*, 407–423. [CrossRef]
34. Nordio, P.L.; Busolin, P. Electron Spin Resonance Line Shapes in Partially Oriented Systems. *J. Chem. Phys.* **1971**, *55*, 5485. [CrossRef]
35. Sebastião, P.J.; Gradišek, A.; Pinto, L.F.V.; Apih, T.; Godinho, M.H.; Vilfan, M. Fast Field-Cycling NMR Relaxometry Study of Chiral and Nonchiral Nematic Liquid Crystals. *J. Phys. Chem. B* **2011**, *115*, 14348. [CrossRef] [PubMed]
36. Torrey, H.C. Nuclear Spin Relaxation by Translational Diffusion. *Phys. Rev.* **1953**, *92*, 962. [CrossRef]

37. Sebastião, P.J. The art of model fitting to experimental results. *Eur. J. Phys.* **2014**, *35*, 015017. [CrossRef]
38. Žumer, S.; Vilfan, M. Theory of nuclear-spin relaxation by translational self-diffusion in liquid-crystals—Nematic phase. *Phys. Rev. A* **1978**, *17*, 424. [CrossRef]
39. Roy, D.; Fragiadakis, D.; Roland, C.M.; Dąbrowski, R.; Dziaduszek, J.; Urban, S. Phase Behavior and Dynamics of a Cholesteric Liquid Crystal. *J. Chem. Phys.* **2014**, *140*, 074502. [CrossRef]

materials

Article

Miscibility Studies of Bismesogen CBnCB Forming Nematic Twist-Bend Phase with Cyanobiphenyls nCB

Marzena Tykarska [1,*], Barbara Klucznik [1], Jerzy Dziaduszek [1] and Stanisław Jóźwiak [2]

[1] Institute of Chemistry, Military University of Technology, ul. gen. Sylwestra Kaliskiego 2, 00-908 Warsaw, Poland; barbara.klucznik@wat.edu.pl (B.K.)
[2] Institute of Materials Science and Engineering, Military University of Technology, ul. gen. Sylwestra Kaliskiego 2, 00-908 Warsaw, Poland; stanislaw.jozwiak@wat.edu.pl
* Correspondence: marzena.tykarska@wat.edu.pl

Abstract: This work aims to determine how the nematic twist-bend phase (N_{TB}) of bismesogens containing two rigid parts of cyanobiphenyls connected with a linking chain containing n = 7, 9, and 11 methylene groups behaves in mixtures with structurally similar cyanobiphenyls nCB, n = 4–12, 14. The whole phase diagrams are presented for the CB7CB-nCB system. For the other systems, CB9CB-nCB and CB11CB-nCB, only curves corresponding to N_{TB}-N phase transition are presented. Based on the temperature-concentration range of the existence of N_{TB} phase, it was established that an increase in the alkyl chain length of CBnCB causes an increase in the stability of the N_{TB} phase. But surprisingly, an increase in the alkyl chain length of nCB compounds does not change the slope of the N_{TB}-N equilibrium line on phase diagrams. It is slightly bigger when the nCB compound has the same length of alkyl chain as the length of the linking group of a bismesogen. XRD studies were carried out for two mixtures.

Keywords: liquid crystals; phase diagrams; miscibility studies; nematic twist-bend phase

1. Introduction

Miscibility studies are often performed because they help in the identification of phases [1–3], allow us to establish the influence of the molecular structure on phase existence [4,5] and to check if compounds are good candidates for mixture components [6,7], because for application purpose mixtures are used. The results of miscibility studies of compounds are presented in phase diagrams, showing phase transition temperatures versus concentration. This is a tool that helps verify the influence of the composition of liquid crystal mixtures on phase situations. The phase may be supported, destabilized, or induced in mixtures [5,7,8]. Even compounds with non-compatible structures are mixed together to check the possibilities of phase formation [9–12].

The liquid crystal nematic phase (N) is formed by elongated molecules with only an orientational order and translating along their long axes. When molecules contain a carbon atom with four different substituents, a chiral nematic phase is formed (N*) because molecules create a macroscopic helix. In 2001, Dozov predicted the possibility of forming a new nematic phase, namely the nematic twist-bend phase (N_{TB}) [13]. Ten years later, the existence of this phase was experimentally confirmed [14]. In contradiction to the chiral nematic phases, in an N_{TB} phase, heliconical structure is formed by achiral molecules and their order is presented in Figure 1.

The first compound for which the N_{TB} phase was confirmed was 1,7-bis(4-cyanobiphenyl-4′-yl)heptane (CB7CB) [14]. This compound consists of two cyanobiphenyl groups linked by an alkyl chain. This compound is the most tested compound forming the N_{TB} phase, e.g., refs. [15–20]. An examination of CBnCB series of bismesogens shows that only compounds containing an odd number of methylene groups (n = 5, 7, 9, 11) in the linking alkyl chain form the N_{TB} phase [21]. More information about the N_{TB} phase can be found, e.g., in refs. [22–26].

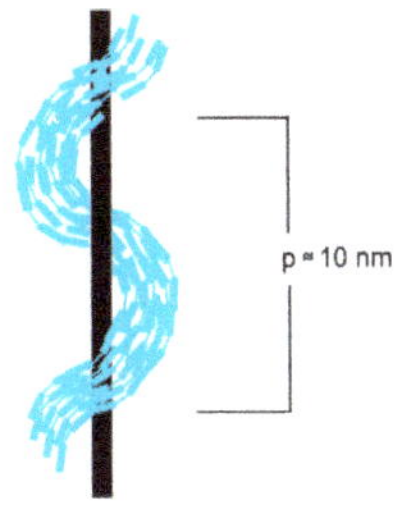

Figure 1. Structure of a twist-bend nematic N_{TB} phase; p–pitch of heliconical structure.

Miscibility studies are performed to test the influence of the molecular structure of different compounds on maintaining the phases. In the literature, we may find examples of miscibility studies of CB7CB with different bismesogens (e.g., refs. [27–29]) and also with monomesogens which are chemically very similar, namely pentylcyanobiphenyl 5CB [30–32] and octylcyanobiphenyl 8CB [33,34]. On phase diagrams, the melting points are not presented so we cannot define the temperature range in which the N_{TB} phase is enantiotropic, as it is essential for the possibilities of making measurements. Recently, we reported the results of miscibility studies of the CB9CB compound with cyanobiphenyls nCB, n = 4–12, 14 [35]. The whole phase diagrams were presented there. The crystal phase is stabilized in mixtures more so than in theoretical predictions.

The aim of this work is to determine the influence of cyanobiphenyls nCB (n = 4–12, 14) on miscibility with CBnCB (n = 7, 9, 11) compounds. The whole phase diagrams are presented for the system CB7CB-nCB, but for CB9CB-nCB and CB11CB-nCB, only lines corresponding to the upper limit of existence of N_{TB} and SmA_d phases are compared. The presented results will enable us to test the influence of the alkyl chain length of tested compounds on the stability of the N_{TB}, SmA_d, and crystal phases. This type of research fits into the topic of determining the influence of the structure of compounds on the formation of liquid crystalline phases.

2. Materials and Methods

The chemical structures of examined compounds CBnCB and nCB are shown in Figure 2. Their phase transition temperatures are given in Table 1. The CBnCB compounds consist of two cyanobiphenyl groups linked with a chain of several methylene groups (n = 7, 9, 11). They form a twist-bend nematic phase N_{TB}. The compounds that are mixed with CBnCB are cyanobiphenyls nCB, which have different alkyl chain lengths n = 4–12 and 14. Compound n = 4 forms a monotropic N phase, compounds n = 5–7 form an enantiotropic N phase, compounds n = 8 and 9 form SmA_d and N phases, and compounds n = 10–12, and 14 form only an SmA_d phase of a dimeric structure (SmA_d [36]).

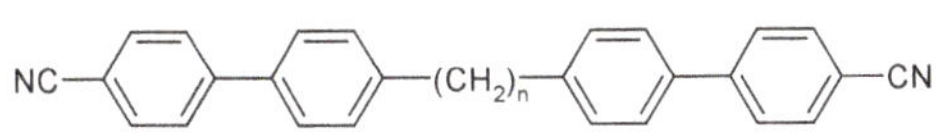

CBnCB, n = 7, 9, 11

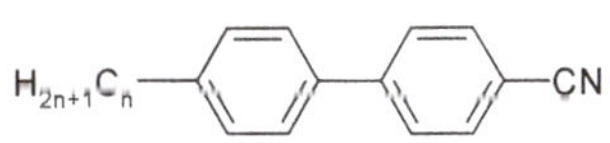

nCB, n = 4–12, 14

Figure 2. Chemical structure of tested compounds.

Table 1. Acronyms and phase transition temperatures (°C) of the tested compounds measured in this research by POM method. Temperatures in brackets denote monotropic transition. These compounds were first reported in refs. presented in the last column.

Acronym	Cr		N_{TB}		SmA_d		N		Iso	Refs.
CB7CB	*	102.9 [1]	*	104.4	-	-	*	117.2	*	[14,37]
CB9CB	*	84.9	*	107.8	-	-	*	123.4	*	[38,39]
CB11CB	*	105.7 [2]	*	108.7	-	-	*	125.4	*	[40]
4CB	*	49.5	-	-	-	-	*	(16.5)	*	[41]
5CB	*	21.6	-	-	-	-	*	39.4	*	[41]
6CB	*	14.8	-	-	-	-	*	29.5	*	[41]
7CB	*	30.5	-	-	-	-	*	43.3	*	[41]
8CB	*	20.3	-	-	*	33.0	*	39.7	*	[41]
9CB	*	43.9	-	-	*	48.4	*	50.3	*	[41]
10CB	*	44.5	-	-	*	51.7	-	-	*	[42]
11CB	*	54.3	-	-	*	58.3	-	-	*	[42]
12CB	*	47.4	-	-	*	59.9	-	-	*	[42]
14CB	*	58.0	-	-	*	63.6	-	-	*	[43]

[1] from DSC: melting of Cr_3 66.9, of Cr_2 102.5, of Cr_1 105.2; N_{TB} 103.1 N 115.7 Iso. [2] from DSC: melting of Cr_2 96.0, of Cr_1 104.7; N_{TB} 108.2 N 124.9 Iso.

The phase transition temperatures were determined from observations of texture changes with a polarizing optical microscope (POM) MP-349 (OPTA-TECH, Warsaw, Poland) equipped with heating stage THMSE6-200 (Linkam Scientific Instruments Ltd., Tadworth, UK) and temperature controller T95-STD (Linkam Scientific Instruments Ltd., Tadworth, UK). The measurements were performed in a heating cycle with a heating rate of 2 K/min near the phase transition region. The results are presented in the form of phase diagrams.

The phase transition enthalpies were measured using differential scanning microcalorimeter DSC 141 (Setaram, Caluire-et-Cuire, France) at both heating and cooling cycles with a scan rate of 2 K/min. This heating and scanning rate is optimal because one smaller than this is too small due to instruments' limitations which cause transitions to be stretched in time, and a bigger one causes a shift in transition temperatures. The X-ray diffractograms were recorded on Ultima IV (Rigaku, Tokyo, Japan) using Co $K\alpha_1$ radiation (λ = 1.789 Å) in the range of 2θ between 5° and 40°. The measurement was carried out in the reflection mode, using a parallel beam (Rigaku patent) with a scanning step of $\Delta 2\theta$ = 0.02° and a scanning speed of 1 °/min using a lamp voltage of 40 kV and a current of 40 mA. The tests were carried out at a constant, small beam incidence angle of 5°. The sample was placed in an open holder made of glass. The test temperature was 23 °C and was controlled using a thermohygrometer placed in the measuring chamber of the diffractometer.

3. Results and Discussion

3.1. Compound CB7CB

When the CB7CB compound was first reported in 1993, its phase transition temperatures were as follows: Cr 101 S 102 N 115 Iso [37]. The N_{TB} phase was incorrectly identified as a smectic phase there because the N_{TB} phase gives textures similar to those observed for the smectic phase. When the existence of the N_{TB} phase was experimentally confirmed in this compound in 2011, the transition temperatures were given as follows: Cr 102 N_{TB} 103 N 116 Iso [14]. In the case of our sample of the CB7CB compound, the phase transition temperatures are shown in Table 1 and Figure 3. The textures of the N and N_{TB} phases are given in Figure 4. We performed a few DSC measurements of the CB7CB compound with a heating/cooling rate of 2 K/min and found that this compound had three different crystal phases. They do not appear in every case. Figure 3a,b present the DSC thermograms obtained by heating and cooling. Curves a–c in Figure 3a obtained from the first heating of different samples show that two crystals Cr_2 and Cr_1 melt at 102.5 and 105.0 °C (peak top of the most separated peaks), respectively. The intensity of these peaks is different depending on the proportion of the existence of both phases in a sample. The d curve obtained from the second heating shows that the Cr_3 phase melts at 66.9 °C and the Cr_1 phase at 105.2 °C. The Scanning Tunneling Microscopy measurements show three different possible arrangements of bismesogenic molecules [44]. It is worth noticing that the sum of enthalpies of these different transitions is comparable (27.35, 27.72, 27.12, and 28.63 kJ/mol from curves a, b, c, and d, respectively). Thus, the disappearance of the crystal phase and formation of the liquid crystal phase require the same amount of energy regardless of the type of crystal phases formed. The N-Iso transition is at 115.8 °C, but in the cooling process, it is a little lower (115.0 °C). This is normal behavior because the transitions occurring during cooling are driven not only thermodynamically (as during heating) but also kinetically. The N_{TB}-N transition is not visible in the DSC presented in Figure 3a, only in Figure 3b during cooling. However, as was mentioned, only transition temperatures measured during heating are characteristic parameters of materials. Such a thermogram, obtained for the CB7CB heated before crystallization appeared (presented in Figure 8a as 0 wt.% of 9CB), shows that the N_{TB}-N transition is at 103.1 °C (the same as in ref [14]).

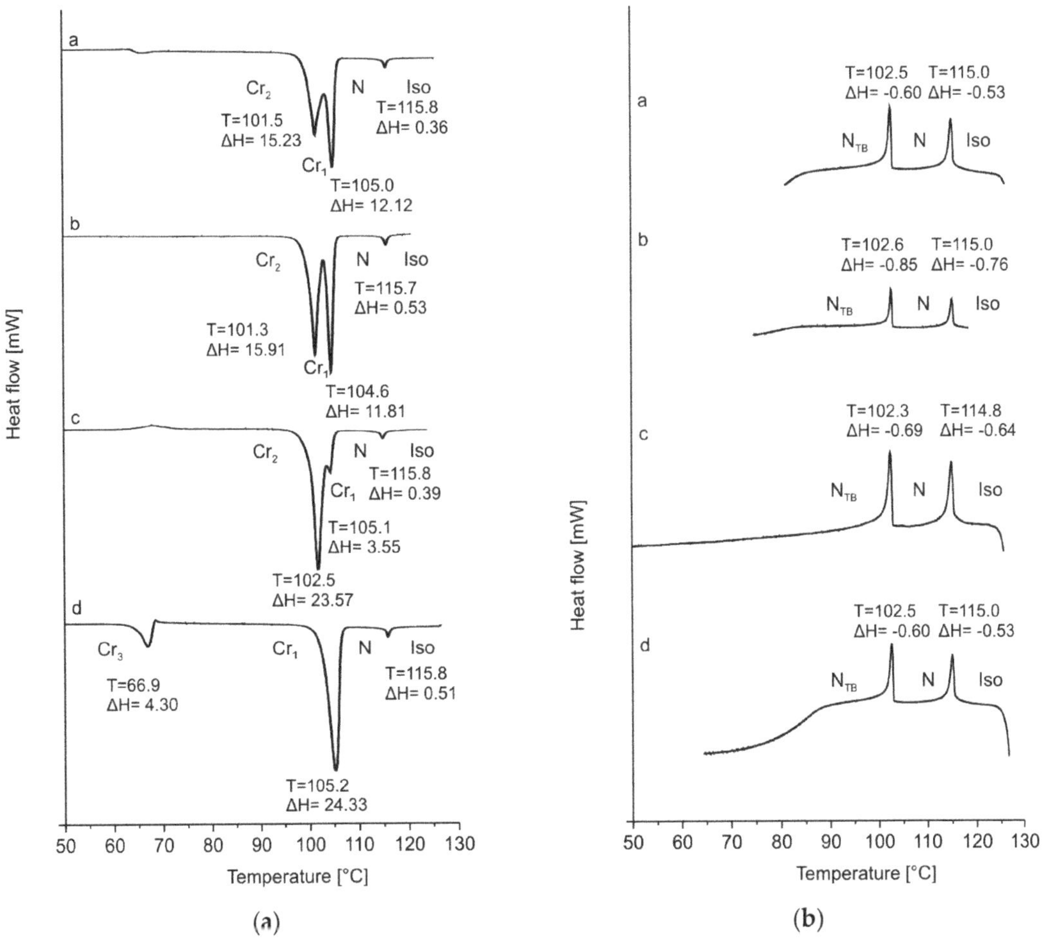

Figure 3. DSC thermograms of CB7CB compound: (**a**) under heating (a–c, the first heating, d, the second heating), (**b**) under cooling (a–c, the first cooling, d, the second cooling); scan rate 2 K/min. Presented T in [°C] and ΔH in [kJ/mol].

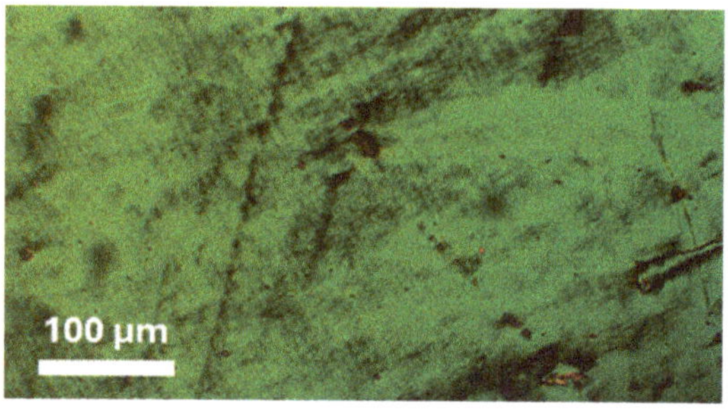

(**a**)

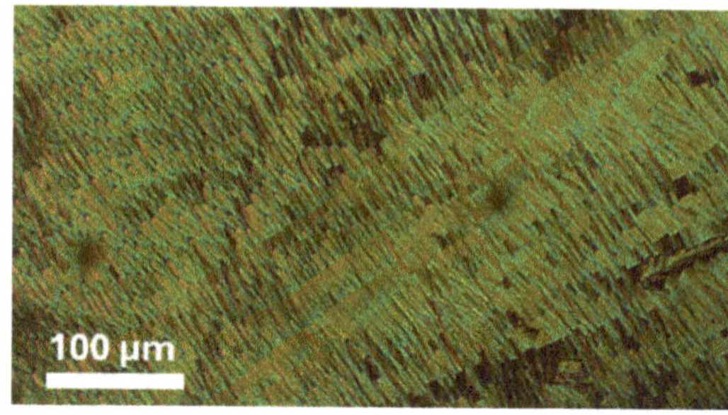

(**b**)

Figure 4. Micrographs of compound CB7CB: (**a**) at 108.8°C in nematic phase; (**b**) at 102.6 °C in twist-bend nematic phase, placed between untreated slides and crossed polarizers.

3.2. Miscibility Studies

3.2.1. Influence of Alkyl Chain Length of nCB on Phase Situation

The CB7CB compound was mixed with nCB compounds (n = 4–12, 14) with a rigid core the same as one part of this bismesogen. The miscibility results are shown in phase diagrams, where phase transition temperatures are marked versus the concentration of the compounds (Figure 5). The concentration is given as a mole fraction; thanks to this, we can compare the results for different homologs. For example, a concentration of 0.5 mole fraction corresponds to a 1:1 ratio of both types of molecules. To achieve such acondition we have to prepare a mixture 40.2 wt.% of 9CB and only 34.1 wt.% of 4CB. All transitions were obtained in a heating cycle, even for the monotropic phase. After cooling, the samples and formation of the N_{TB} phase, before it was crystallized, it was heated.

The situation with **clearing transition** is different and depends on the kind of mesophases formed by cyanobiphenyls. When compounds nCB form a nematic phase, whether monotropic (n = 4) or enantiotropic (n = 5–9), it mixes additively with a nematic phase of CB7CB, and a biphase region (N + Iso) is observed within a temperature range with a maximum of 3 degrees. When compounds nCB form only the smectic A_d phase (n = 10–12, 14), the nematic and SmA_d phases undergo an eutectic transition. The temperature range of the biphase regions is too small to be well presented.

The **crystal phase** melts according to a typical eutectic transition. The temperature of an eutectic mixture is marked with a short horizontal line. In all cases, an eutectic concentration is close to the nCB compound. For systems containing nCB n = 9, 11, 14, the crystallization was in the whole concentration range, but the other systems did not crystallize with a higher concentration of cyanobiphenyls. This is why the melting temperatures are not marked in a full concentration range. A comparison of melting temperature versus the concentration of compounds is presented in Figure 6. The calculated melting curve according to Le Chatelier, Schröder, and Van Laar (CSL) Equation (1), satisfying condition (2), is marked with a dashed line.

$$\ln x_k = \frac{\Delta H_m^k}{R}\left(\frac{1}{T} - \frac{1}{T_m^k}\right), \quad (1)$$

$$\sum_{k=1}^{n} x_k = 1 \quad (2)$$

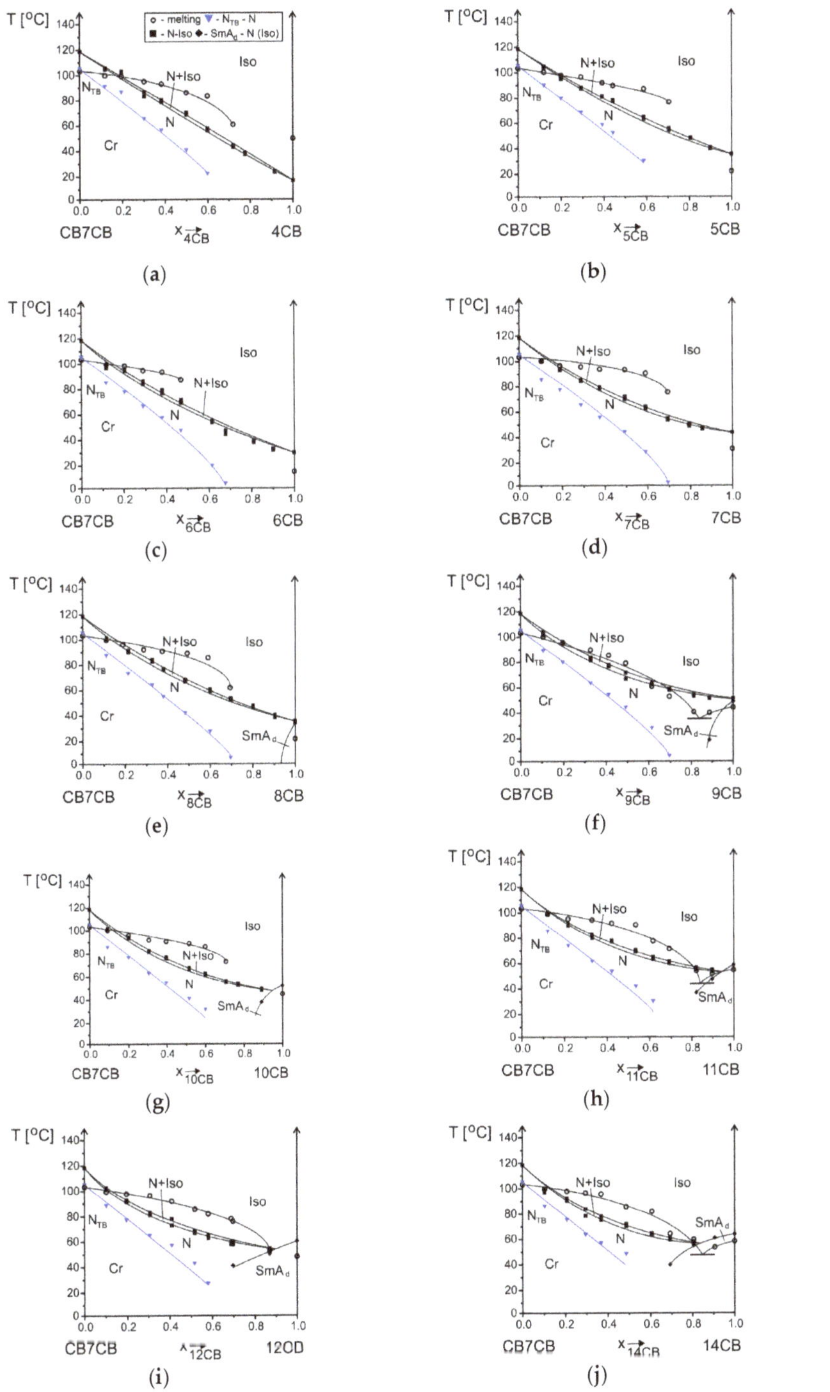

Figure 5. Phase diagrams presented for the CB7CB-nCB systems, n = 4 (**a**), 5 (**b**), 6 (**c**), 7 (**d**), 8 (**e**), 9 (**f**), 10 (**g**), 11 (**h**), 12 (**i**), 14 (**j**). Transition temperatures measured upon heating. Legend in (**a**) common for all graphs.

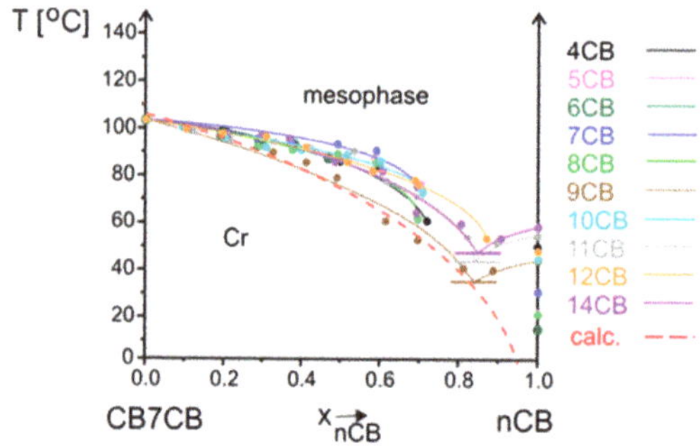

Figure 6. Melting temperatures for CB7CB-nCB system (n = 4–12, 14) versus concentration x (mole fraction). The results of the CSL calculations are marked with a dashed line.

The measured melting temperature for most systems is observed to be higher than calculated. The crystal phase is more stable than the theory predicts; thus, intermolecular interactions between molecules in mixtures are stronger than in single compounds. The biggest difference is for the system with 7CB compound, which only slightly decreases the melting temperature, and up to 0.6 mole fraction, it is similar to the pure CB7CB compound. The $\mathbf{N_{TB}}$ **phase** is monotropic; thus, it can be reached during the cooling process from a nematic phase. However, the phase transition temperature measured upon cooling is not a characteristic parameter of a material. This is why, after cooling the sample at least 5 degrees below this transition and before crystallization started, the temperature was increased and N_{TB}-N transition temperature was measured upon heating. The lines corresponding to the equilibrium between these two phases versus concentration are compared in one summary phase diagram presented in Figure 7a. The stability of the N_{TB} phase of the CB7CB compound decreases in mixtures upon an increase in the concentration of nCB compounds. These lines have the same slope. The same observation was made for systems where a longer bismesogen CB9CB was doped with cyanobiphenyls of different alkyl chain lengths [35]. The dependence of the temperature of maximum stability of the N_{TB} phase upon the alkyl chain length of nCB compounds, for concentration x = 0.3 mole fraction of nCB, is presented in Figure 7b. Even though the values are within 4 degrees, the odd–even effect in the results can be noticed. The maximum temperature of the N_{TB}-N transition is for the system where the 7CB compound is added to CB7CB. For systems CB9CB-nCB, the maximum was observed for 9CB [35]. The mentioned observations show that the most stable N_{TB} phase is obtained when the alkyl chain length of cyanobiphenyl has the same length as the linking alkyl chain of a bismesogen. Molecules can fit better into the structure formed by a bismesogen. The maximum concentration of the existence of the N_{TB} phase is mostly up to a 0.6–0.7 mole fraction of nCB. The exact border concentrations of N_{TB} phase existence are difficult to establish because mixtures crystalize before the N_{TB} phase appears. The N_{TB}-N transition on DSC thermograms is visible only upon second heating. In Figure 8, an example of DSC thermograms for the system CB7CB-9CB are shown. For mixtures containing a 0.5 and 0.6 mole fraction of 9CB, the N_{TB}-N transition is not visible because the N_{TB} phase supercools very easily and does not appear under conditions of the DSC measurement. But it was visible during POM measurements; see micrographs of the textures recorded for three mixtures of CB7CB-9CB containing 0.3, 0.4, and 0.5 mole fractions of 9CB given in Figure 9. The change in the enthalpy of the N_{TB}-N transition versus concentration shows that it decreases linearly and, for mixtures of 0.5 and 0.6 of 9CB, should be smaller than 0.1 kJ/mol. Mixtures containing CB7CB supercool to low temperatures, the easiest of all mixtures containing CBnCB compounds. This is why it is possible to measure the properties of the monotropic N_{TB} phase in a broad temperature–concentration range.

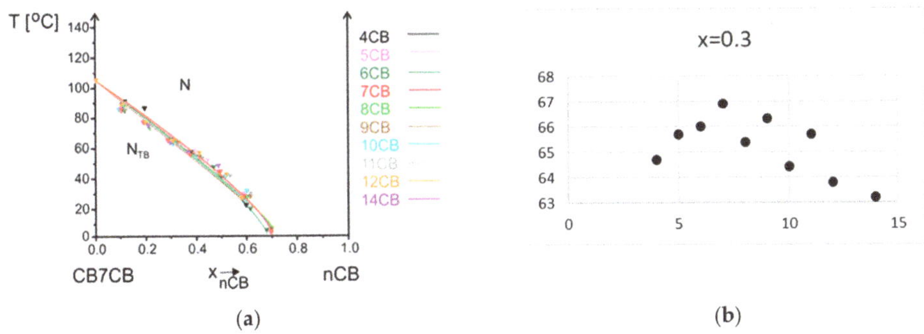

Figure 7. Phase transition N_{TB}-N temperatures for CB7CB-nCB system (n = 4–12, 14) versus (**a**) concentration x (mole fraction), (**b**) alkyl chain length n for set concentration x = 0.3 mole fraction of nCB.

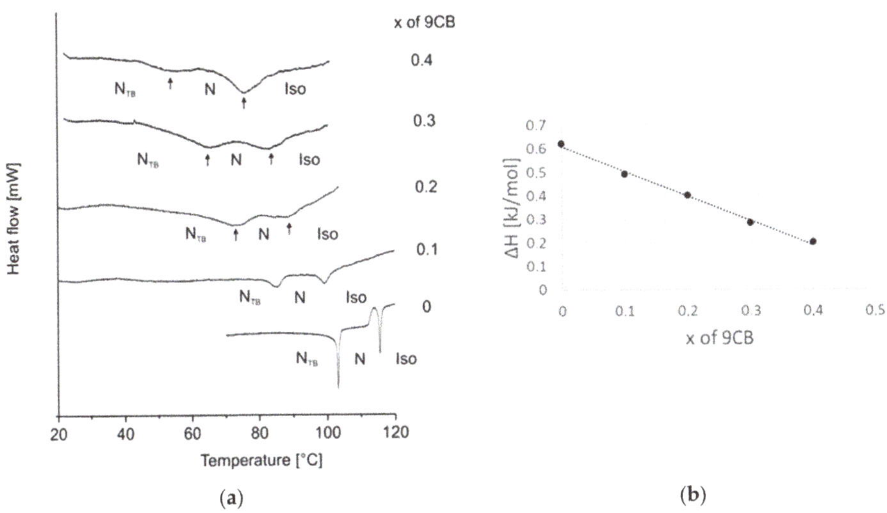

Figure 8. (**a**) DSC thermograms of CB7CB-9CB system; second heating started at temperatures before crystallization of first cooling was started. Concentration "0" corresponds to pure CB7CB; (**b**) changes in N_{TB}-N transition enthalpies versus concentration in this system.

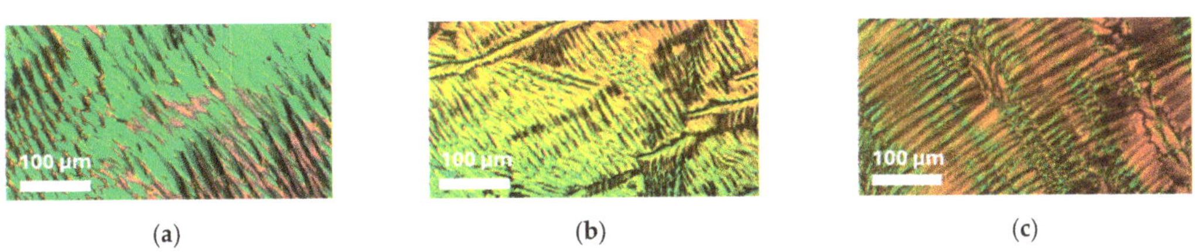

Figure 9. Micrographs of twist-bend nematic phase observed for mixtures CB7CB-9CB containing (**a**) 0.3 at 57.3 °C; (**b**) 0.4 at 49.8 °C; (**c**) 0.5 at 44.2 °C mole fractions of 9CB. Placed between untreated slides and crossed polarizers.

Regarding the **SmA$_d$ phase** which is formed by nCB with n = 8–12, 14, the maximum temperature of the existence of this phase increases as the alkyl chain length of the cyanobiphenyls increases (Table 1). The same is observed in mixtures with bismesogens

(Figure 10a). The range of the concentration of the existence of this phase increases and the maximum temperature is not destabilized so rapidly. However, it is hard to establish how close the N_{TB} phase is to the SmA_d phase because the samples crystalize before reaching temperatures in which it may be observed (Figure 5e–j). The phase diagram presented by Aouini et al. [34] for the CB7CB-8CB system was enriched with the results of dielectric measurements. The equilibrium lines N_{TB}-N and SmA_d-N were very close to each other, but the authors could not confirm the transition between the N_{TB} and SmA_d phases.

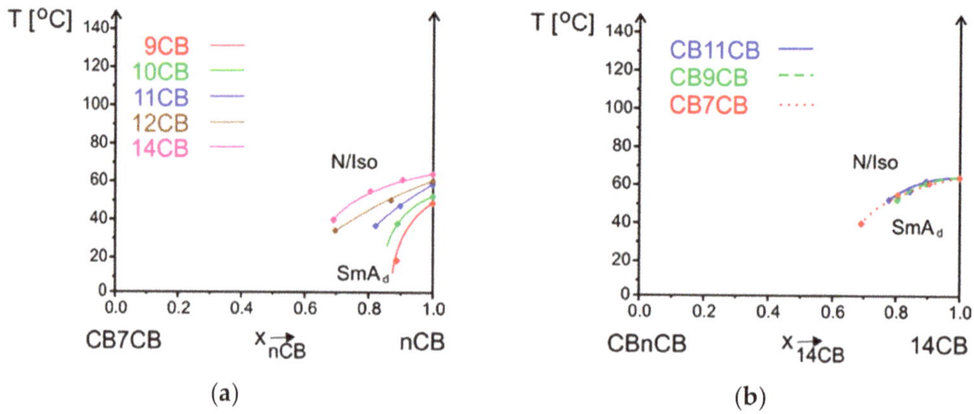

Figure 10. Comparison of lines corresponding to phase equilibrium SmA_d-N/Iso for systems (**a**) CB7CB-nCB (n = 9–12), 14; (**b**) CBnCB-14CB (n = 7, 9, 11).

3.2.2. Influence of Alkyl Chain Length of CBnCB on Phase Situation

The influence of the alkyl chain length of bismesogens CB7CB, CB9CB, and CB11CB in mixtures with nCB on N_{TB}-N transition temperature was determined (Figure 11). In all systems, an increase in the number of methylene groups (n = 7, 9, 11) in the alkyl chain of bismesogen causes an increase in the N_{TB}-N transition temperatures. The maximum concentration for which this phase transition is observed is between the 0.5 and 0.8 mole ratio of nCB. We may assume that the concentration limit is set at a satisfactory level for the systems where the line corresponding to N_{TB}-N phase equilibrium reached 0 °C. In all other cases, the samples were crystallized before the monotropic phase was reached; thus, this transition was not possible to be measured for a bigger concentration of cyanobiphenyl.

The influence of the linking chain length on the SmA_d phase is presented in Figure 10b, where lines corresponding to the phase equilibrium SmA_d-N/Iso are presented. Their slopes are the same, which may suggest that the mesogenic units build in the smectic layer structure regardless of the spacing length. In particular, bismesogens can form the SmA phase when the linking group is sufficiently long [22].

3.3. *XRD Studies of Mixtures CB7CB with 9CB and 4CB*

The X-ray diffraction results for pure CB7CB compounds are already reported in the literature [14]. The two reflexes 4.5 Å and 12 Å are present on diffractograms. Taking into account that the molecular length of this molecule is 26 Å, the former reflex corresponds to the average distance between molecules and the latter indicates the intercalated structure [22]. The rigid arms of bismesogens are ordered side by side but in such a way that one arm is adjacent to one and the other to another bismesogen.

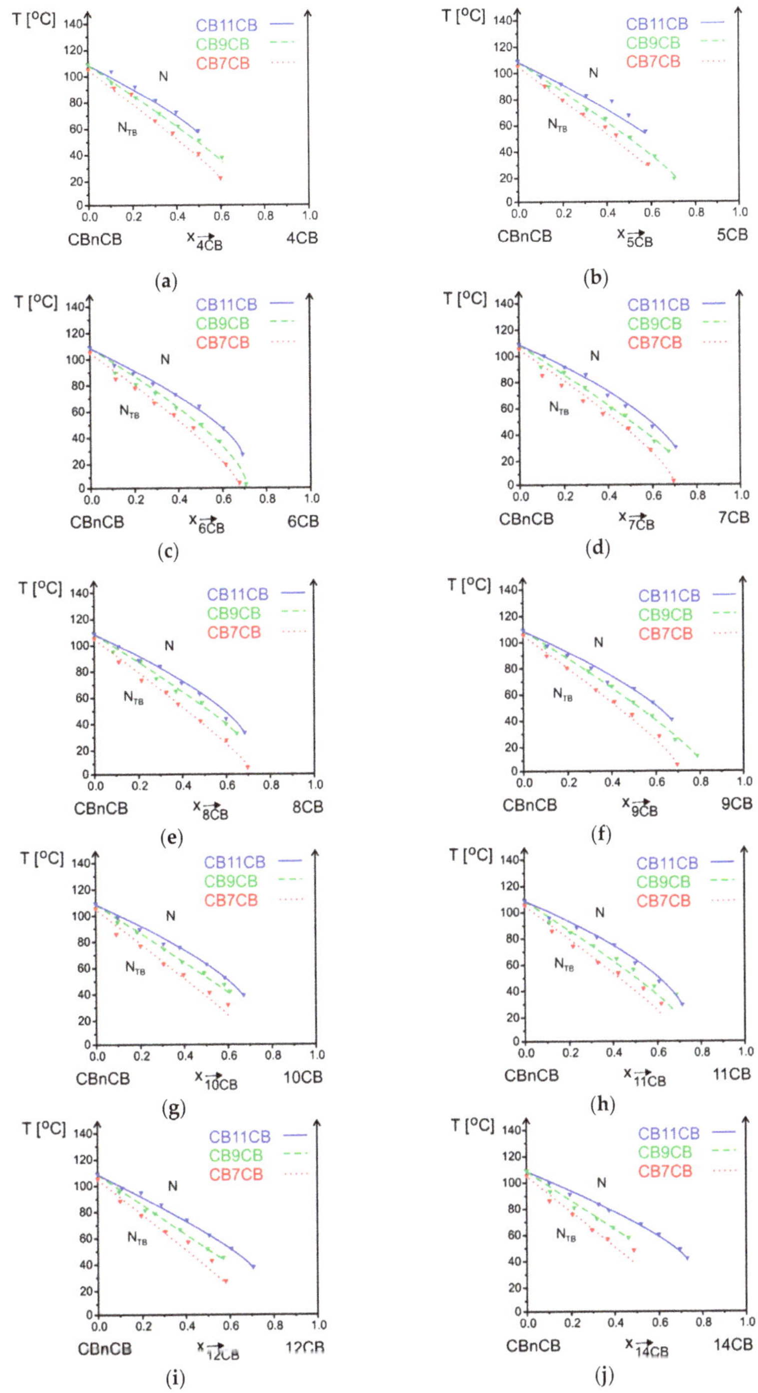

Figure 11. Influence of the length of the alkyl chain of bismesogen CBnCB n = 7, 9, 11 in mixtures with nCB (n = 4 (**a**), 5 (**b**), 6 (**c**), 7 (**d**), 8 (**e**), 9 (**f**), 10 (**g**), 11 (**h**), 12 (**i**), 14 (**j**)) on phase transition N_{TB}-N temperatures versus concentration x (mole fraction).

We studied two mixtures, CB7CB-9CB and CB7CB-4CB, both of which contained a 0.5 mole fraction of cyanobiphenyl. The results of the studies are presented in Figure 12. For both systems, reflexes corresponding to the average separation of molecules (4.4 Å) are observed. The correlation length calculated from the Scherrer equation is 13.5 Å for this signal, which corresponds to three neighboring molecules. For a mixture containing CB7CB with 9CB, the second reflex is exactly of the same value as for pure bismesogen and is close to half of the length of the CB7CB molecule (12Å). This suggests that molecules with a longer tail adopt the structure of the CB7CB host molecules. The calculated correlation length is 38.2 Å, three times bigger than the measured distance. For a mixture containing the 4CB compound, the small angle reflex has two maxima. The reflex giving a distance of 9.3 Å corresponds to the length of the cyanobiphenyl group of both compounds. The next small reflex giving a distance of 7.0 Å corresponds to the length of the linking group of CB7CB. This may suggest that the presence of 4CB molecules causes the segregation of rigid and flexible parts, even more than in pure bismesogen.

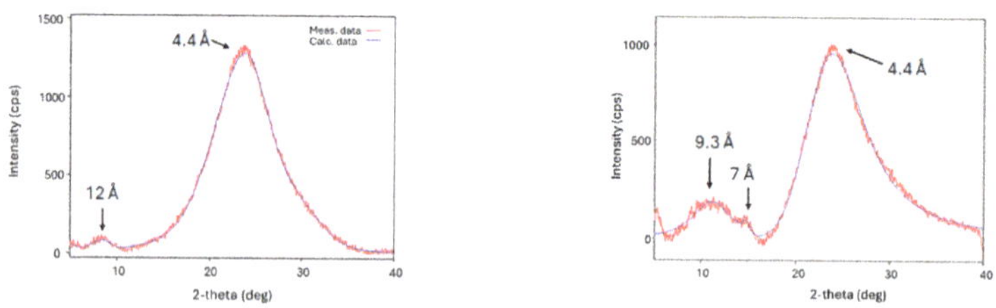

(**a**) (**b**)

Figure 12. X-ray diffractograms for mixtures: (**a**) CB7CB-9CB; (**b**) CB7CB-4CB. Both mixtures contain 0.5 mole fraction of cyanobiphenyl. Measurement temperature is 23 °C.

4. Conclusions

Ten homologs of cyanobiphenyls nCB (n = 4–12, 14) were added to structurally similar bismesogens CBnCB (n = 7, 9, 11), forming a nematic twist-bend phase. They are good candidates as components for mixtures showing this phase because they support this phase in a broad concentration range (from 0.5 to 0.8 mole fraction). It is worth noticing that the slope of the equilibrium curves N_{TB}-N is the same regardless of the length of cyanobiphenyl. Slightly bigger N_{TB} phase stability occurs when the alkyl chain length of cyanobiphenyl is of the same length as the linking alkyl chain of bismesogen CBnCB. On the other side, an increase in the length of the linking alkyl chain of a bismesogen causes an increase in the N_{TB} phase stability both for pure compounds and their mixtures with cyanobiphenyls. The biggest problem with the N_{TB} phase is that this phase is mostly monotropic. Unfortunately, the melting point is not lowered by nCB compounds as much as predicted. Because the rigid parts of bismesogen and monomesogen are the same, molecules in deficiency incorporate into the arrangement of molecules in excess. This is seen both for the nCB built in the structure of the N_{TB} of CBnCB and for the CBnCB built in the structure of SmA_d of nCB. The slope of equilibrium curves on the phase diagrams is the same, regardless of the alkyl chain length. When molecules of cyanobiphenyls are short, they cause the segregation of rigid and flexible parts (mixture CB7CB-4CB), in contrast to the CB7CB-9CB mixture, where longer molecules of cyanobiphenyls better fit the structure.

It is expected that the N_{TB} phase will be obtained at room temperature as an enantiotropic phase. New chemical structures can be searched for, but it is much easier to do this by formulating multicomponent mixtures in which the melting point is lowered. Miscibility studies allow for the determination of which compounds are favorable for the formation of such mixtures. Further studies will be conducted on the influence of the chemical structure of monomesogens, e.g., the core structure or the type of terminal groups, on the behavior of the N_{TB} phase in mixtures.

Author Contributions: Conceptualization, M.T.; validation, M.T., B.K. and S.J.; investigation, B.K., J.D. and S.J.; writing—original draft preparation, M.T.; writing—review and editing, M.T.; visualization, M.T. and B.K.; supervision, M.T. All authors have read and agreed to the published version of the manuscript.

Funding: This research was funded by WOJSKOWA AKADEMIA TECHNICZNA, grant number UGB 22-720.

Institutional Review Board Statement: Not applicable.

Informed Consent Statement: Not applicable.

Data Availability Statement: Data are contained within the article.

Conflicts of Interest: The authors declare no conflicts of interest. The funders had no role in the design of the study; in the collection, analyses, or interpretation of data; in the writing of the manuscript; or in the decision to publish the results.

References

1. Czupryński, K.; Przedmojski, J.; Dąbrowski, R.; Brodzik, M. Reinvestigation of mesomorphic properties of 4,4′-dialkylbiphenyl. New orthogonal phases. *Mol. Cryst. Liq. Cryst.* **1997**, *299*, 503–508. [CrossRef]
2. Bubnov, A.; Tykarska, M.; Hamplová, V.; Kurp, K. Tuning the phase diagrams: The miscibility studies of multilactate liquid crystalline compounds. *Phase Transit.* **2016**, *89*, 885–893. [CrossRef]
3. Kurp, K.; Tykarska, M. Liquid-crystalline properties of laterally fluorinated alkyl 4″-alkylterphenyl-4-yl carbonates and their mixtures. *Liq. Cryst.* **2016**, *43*, 1359–1364. [CrossRef]
4. Tykarska, M.; Skrzypek, K. The comparison of the ability for induction of antiferroelectric phase of cyanoterminated compounds by means of miscibility studies. *Phase Transit.* **2007**, *80*, 203–207. [CrossRef]
5. Zieja, P.; Tykarska, M.; Morawiak, P.; Piecek, W. Influence of the molecular structure of calamitic compounds with trifluoromethoxy terminal group on the induction of the smectic A phase. Part I. *Liq. Cryst.* **2023**, *50*, 363–378. [CrossRef]
6. Tykarska, M.; Kurp, K.; Zieja, P.; Herman, J.; Stulov, S.; Bubnov, A. New quaterphenyls laterally substituted by methyl group and their influence on the self-assembling behaviour of ferroelectric bicomponent mixtures. *Liq. Cryst.* **2022**, *49*, 821–835. [CrossRef]
7. Tykarska, M.; Kurp, K.; Mironov, S.; Rychłowicz, N.; Karcz, J.; Dziaduszek, J.; Kula, P.; Stulov, S.; Bubnov, A. Tuning of self-organizing and electro-optical behaviour for orthoconic ferroelectric liquid crystal by non-chiral dopants. *J. Mol. Liq.* **2024**, *409*, 125426. [CrossRef]
8. Dąbrowski, R.; Czupryński, K. Induced smectic and nematic phases and re-entrant phenomena. In *Modern Topics in Liquid Crystals: From Neutron Scattering to Ferroelectricity*; Buka, A., Ed.; World Scientific Pub. Co. Inc.: Singapore, 1993; pp. 125–160.
9. Nakata, M.; Takanishi, Y.; Watanabe, J.; Takezoe, H. Blue phases induced by doping chiral nematic liquid crystals with nonchiral molecules. *Phys. Rev. E* **2003**, *68*, 041710. [CrossRef]
10. Taushanoff, S.; Van Le, K.; Williams, J.; Twieg, R.J.; Sadashiva, B.K.; Takezoe, H.; Jakli, A. Stable amorphous blue phase of bent-core nematic liquid crystals doped with a chiral material. *J. Mater. Chem.* **2010**, *20*, 5893–5898. [CrossRef]
11. Bubnov, A.; Hamplová, V.; Kašpar, M.; Vajda, A.; Stojanović, M.; Obadović, D.Ž.; Éber, N.; Fodor-Csorba, K. Thermal analysis of binary liquid crystalline mixtures. System of bent core and calamitic molecules. *J. Therm. Anal. Calorim.* **2007**, *90*, 431–441. [CrossRef]
12. Chakraborty, S.; Das, M.K.; Bubnov, A.; Weissflog, W.; Węgłowska, D.; Dąbrowski, R. Induced frustrated twist grain boundary liquid crystalline phases in binary mixtures of achiral hockey stick-shaped and chiral rod-like materials. *J. Mater. Chem. C* **2019**, *7*, 10530–10543.
13. Dozov, I. On the spontaneous symmetry breaking in the mesophases of achiral banana-shaped molecules. *Europhys. Lett.* **2001**, *56*, 247–253. [CrossRef]
14. Cestari, M.; Diez-Berart, S.; Dunmur, D.A.; Ferrarini, A.; de la Fuente, M.R.; Jackson, D.J.B.; López, D.O.; Luckhurst, G.R.; Pérez-Jubindo, M.A.; Richardson, R.M.; et al. Phase behavior and properties of the liquid-crystal dimer 1,7-bis(4-cyanobiphenyl-4-yl)heptane: A twist-bend nematic liquid crystal. *Phys. Rev. E* **2011**, *84*, 031704.
15. Beguin, L.; Emsley, J.W.; Lelli, M.; Lesage, A.; Luckhurst, G.R.; Timimi, B.A.; Zimmermann, H. The Chirality of a Twist–Bend Nematic Phase Identified by NMR Spectroscopy. *J. Phys. Chem. B* **2012**, *116*, 7940–7951. [PubMed]
16. López, D.O.; Sebastian, N.; de la Fuente, M.R.; Martínez-García, J.C.; Salud, J.; Pérez-Jubindo, M.A.; Diez-Berart, S.; Dunmur, D.A.; Luckhurst, G.R. Disentangling molecular motions involved in the glass transition of a twist-bend nematic liquid crystal through dielectric studies. *J. Chem. Phys.* **2012**, *137*, 034502.
17. Jokisaari, J.P.; Luckhurst, G.R.; Timimi, B.A.; Zhu, J.; Zimmermann, H. Twist bend nematic phase of the liquid crystal dimer CB7CB: Orientational order and conical angle determined by 129Xe and 2H NMR spectroscopy. *Liq. Cryst.* **2015**, *42*, 708–721.
18. Zhu, C.; Tuchband, M.R.; Young, A.; Shuai, M.; Scarbrough, A.; Walba, D.M.; Maclennan, J.E.; Wang, C.; Hexemer, A.; Clark, N.A. Resonant Carbon K-Edge Soft X-Ray Scattering from Lattice-Free Heliconical Molecular Ordering: Soft Dilative Elasticity of the Twist-Bend Liquid Crystal Phase. *Phys. Rev. Lett.* **2016**, *116*, 147803.

19. Babakhanova, G.; Parsouzi, Z.; Paladugu, S.; Wang, H.; Nastishin, Y.A.; Shiyanovskii, S.V.; Sprunt, S.; Lavrentovich, O.D. Elastic and viscous properties of the nematic dimer CB7CB. *Phys. Rev. E* **2017**, *96*, 062704.
20. Krishnamurthy, K.S.; Kanakala, M.B.; Yelamaggad, C.V.; Kleman, M. Instabilities in the electric Freedericksz state of the twist-bend nematic liquid crystal CB7CB. *Soft Matter* **2018**, *14*, 5393.
21. Paterson, D.A.; Abberley, J.P.; Harrison, W.T.A.; Storey, J.M.D.; Imrie, C.T. Cyanobiphenyl-based liquid crystal dimers and the twist-bend nematic phase. *Liq. Cryst.* **2017**, *44*, 127–146.
22. Imrie, C.T.; Walker, R.; Storey, J.M.D.; Górecka, E.; Pociecha, D. Liquid Crystal Dimers and Smectic Phases from the Intercalated to the Twist-Bend. *Crystals* **2022**, *12*, 1245. [CrossRef]
23. Mandle, R.J. A Ten-Year Perspective on Twist-Bend Nematic Materials. *Molecules* **2022**, *27*, 2689. [CrossRef] [PubMed]
24. Dunmur, D.A. Anatomy of a Discovery: The Twist–Bend Nematic Phase. *Crystals* **2022**, *12*, 309. [CrossRef]
25. Kumar, M.P.; Karcz, J.; Kula, P.; Dhara, S. Dynamics of a sheared twist-bend nematic liquid crystal. *Phys. Rev. Mat.* **2021**, *5*, 115605. [CrossRef]
26. Merkel, K.; Loska, B.; Arakawa, Y.; Mehl, G.H.; Karcz, J.; Kocot, A. How do intermolecular interactions evolve at the Nematic to Twist–Bent phase transition? *Int. J. Mol. Sci.* **2022**, *23*, 11018. [CrossRef] [PubMed]
27. Lu, Z.; Henderson, P.A.; Paterson, B.J.A.; Imrie, C.T. Liquid crystal dimers and the twist-bend nematic phase. The preparation and characterization of the α,ω-bis(4-cyanobiphenyl-4′-yl) alkanedioates. *Liq. Cryst.* **2014**, *41*, 471–483. [CrossRef]
28. Robles-Hernandez, B.; Sebastian, N.; Salud, J.; Diez-Berart, S.; Dunmur, D.A.; Luckhurst, G.R.; López, D.O.; de la Fuente, M.R. Molecular dynamics of a binary mixture of twist-bend nematic liquid crystal dimers studied by dielectric spectroscopy. *Phys. Rev. E* **2016**, *93*, 062705.
29. Paterson, D.A.; Gao, M.; Kim, Y.-K.; Jamali, A.; Finley, K.L.; Robles-Hernandez, B.; Diez-Berart, S.; Salud, J.; de la Fuente, M.R.; Timimi, B.A.; et al. Understanding the twist-bend nematic phase: The characterisation of 1-(4-cyanobiphenyl-40-yloxy)-6-(4-cyanobiphenyl-40-yl)hexane (CB6OCB) and comparison with CB7CB. *Soft Matter* **2016**, *12*, 6827. [CrossRef]
30. Tuchband, M.R.; Shuai, M.; Graber, K.A.; Chen, D.; Radzihovsky, L.; Klittnick, A.; Foley, L.; Scarbrough, A.; Porada, J.H.; Moran, M.; et al. The twist-bend nematic phase of bent mesogenic dimer CB7CB and its mixtures. *arXiv* **2015**, arXiv:1511.07523.
31. Trbojevic, N.; Read, D.J.; Nagaraj, M. Dielectric properties of liquid crystalline dimer mixtures exhibiting the nematic and twist-bend nematic phases. *Phys. Rev. E* **2017**, *96*, 052703.
32. Zavvou, E.E.; Ramou, E.; Ahmed, Z.; Welch, C.; Mehl, G.H.; Vanakaras, A.G.; Karahaliou, P.K. Dipole-dipole correlations in the nematic phases of symmetric cyanobiphenyl dimers and their binary mixtures with 5CB. *Soft Matter* **2023**, *19*, 9224. [CrossRef] [PubMed]
33. Oswald, P. Surface-field-induced heliconical instability in the cholesteric phase of a mixture of a flexible dimer (CB7CB) and a rodlike molecule (8CB). *Phys. Rev. E* **2022**, *105*, 024704. [CrossRef] [PubMed]
34. Aouini, A.; Nobili, M.; Chauveau, E.; Dieudonné-George, P.; Damême, G.; Stoenescu, D.; Dozov, I.; Blanc, C. Chemical-Physical Characterisation of a Binary Mixture of a Twist Bend Nematic Liquid Crystal with a Smectogen. *Crystals* **2020**, *10*, 1110. [CrossRef]
35. Klucznik, B.; Tykarska, M. Miscibility studies of bimesogen CB9CB with members of homologous series nCB. *Phase Transit.* 2024, *submitted*.
36. Gray, G.W.; Lydon, J.E. New type of smectic mesophase? *Nature* **1974**, *252*, 221–222. [CrossRef]
37. Barnes, P.J.; Douglass, A.G.; Heeks, S.K.; Luckhurst, G.R. An enhanced odd-even effect of liquid crystal dimers Orientational order in the α,ω-bis(4′-cyanobiphenyl-4-yl)alkanes. *Liq. Cryst.* **1993**, *13*, 603–613. [CrossRef]
38. Robles-Hernandez, B.; Sebastian, N.; de la Fuente, M.R.; López, D.O.; Diez-Berart, S.; Salud, J.; Blanca Ros, M.; Dunmur, D.A.; Luckhurst, G.R.; Timimi, B.A. Twist, tilt, and orientational order at the nematic to twist-bend nematic phase transition of 1″,9″-bis(4-cyanobiphenyl-4′-yl) nonane: A dielectric, 2H NMR, and calorimetric study. *Phys. Rev. E* **2015**, *92*, 062505. [CrossRef] [PubMed]
39. Tripathi, C.S.P.; Losada-Perez, P.; Glorieux, C.; Kohlmeier, A.; Tamba, M.G.; Mehl, G.H. Nematic-nematic phase transition in the liquid crystal dimer CBC9CB and its mixtures with 5CB: A high-resolution adiabatic scanning calorimetric study. *Phys. Rev. E* **2011**, *84*, 041707. [CrossRef]
40. Mandle, R.J.; Davis, E.J.; Archbold, C.T.; Cowling, S.J.; Goodby, J.W. Microscopy studies of the nematic N_{TB} phase of 1,11-di-(1″-cyanobiphenyl-4-yl)undecane. *J. Mater. Chem. C* **2014**, *2*, 556–566.
41. Gray, G.W.; Harrison, K.J.; Nash, J.A. New family of nematic liquid crystals for displays. *Electron. Lett.* **1973**, *9*, 130–131. [CrossRef]
42. Urban, S.; Massalska-Arodź, M.; Würflinger, A.; Dąbrowski, R. Pressure-temperature phase diagrams of smectogenic 4′-alkyl-4-cyanobiphenyl (9CB, 10CB, 11CB and 12CB). *Liq. Cryst.* **2003**, *30*, 313–318. [CrossRef]
43. Urban, S.; Würflinger, A.; Gestblom, B.; Dąbrowski, R.; Przedmojski, J. DTA, X-ray and dielectric relaxation studies of 14CB at atmospheric and elevated pressures. *Liq. Cryst.* **2003**, *30*, 305–311. [CrossRef]
44. Krzyżewska, K.; Jaroch, T.; Maranda-Niedbała, A.; Pociecha, D.; Górecka, E.; Ahmed, Z.; Welch, C.; Mehl, G.H.; Proń, A.; Nowakowski, R. Supramolecular organization of liquid-crystal dimers-bis-cyanobiphenyl alkanes on HOPG by scanning tunneling microscopy. *Nanoscale* **2018**, *10*, 16201–16210. [CrossRef] [PubMed]

materials

MDPI

Article

Synthesis and Properties of Highly Tilted Antiferroelectric Liquid Crystalline (*R*) Enantiomers

Magdalena Urbańska [1,*], Monika Zając [1], Michał Czerwiński [1], Przemysław Morawiak [2], Alexej Bubnov [3] and Aleksandra Deptuch [4]

1 Institute of Chemistry, Military University of Technology, ul. Sylwestra Kaliskiego 2, 00-908 Warsaw, Poland; monika.zajac@wat.edu.pl (M.Z.); michal.czerwinski@wat.edu.pl (M.C.)
2 Institute of Applied Physics, Military University of Technology, ul. Sylwestra Kaliskiego 2, 00-908 Warsaw, Poland; przemyslaw.morawiak@wat.edu.pl
3 Institute of Physics of the Czech Academy of Sciences, Na Slovance 1999/2, 18200 Prague, Czech Republic; bubnov@fzu.cz
4 Institute of Nuclear Physics Polish Academy of Sciences, Radzikowskiego 152, 31-342 Krakow, Poland; aleksandra.deptuch@ifj.edu.pl
* Correspondence: magdalena.urbanska@wat.edu.pl; Tel.:+48-261-83-75-49

Abstract: This work reports the synthesis method and various properties of four rod-like antiferroelectric (*R*) laterally substituted enantiomers, with or without fluorine atoms used as substituents in the benzene ring. The influence of fluorine substitution on the mesophase temperature range was determined. The synthesized compounds are three-ring rod-like smectics with a chiral center based on (*R*)-(−)-2-octanol. Their chemical and optical purity was checked using high-performance liquid chromatography (HPLC). Two newly synthesized enantiomers and three previously reported (*R*) enantiomers were used to formulate two antiferroelectric mixtures. The mesomorphic behavior was characterized by polarizing optical microscopy, differential scanning calorimetry, and X-ray diffraction (XRD). The helical pitch and tilt angle measurements were done using the selective light reflection phenomenon and the electro-optical method, respectively. All the enantiomers exhibit a wide temperature range of the antiferroelectric phase, with a high tilt angle. Furthermore, the enantiomer with lateral fluorine substitution in the ortho position has a very long helical pitch (more than 2.0 μm), relatively low enthalpy of melting point, and a tilt angle close to 45 degrees. The designed (*R*) enantiomers can be useful for formulating eutectic mixtures for further use in various devices, including photonics and optoelectronics.

Keywords: (*R*) enantiomer; synthesis; HPLC; SmC_A* phase; helical pitch; tilt angle; eutectic mixtures; X-ray diffraction

1. Introduction

In smectic liquid crystalline (LC) phases, the molecules exhibit translational order in a layered structure, with at least short-range positional order within layers [1,2]. The formation of chiral smectic LC phases results from the ordered propagation of the tilt angle, layer by layer, so the assembly of molecular units adopts the helical arrangement on a macroscopic level [3,4]. Since the first reports on chiral synclinic (ferroelectric) [5] and chiral anticlinic (antiferroelectric) [6] LCs appeared, significant research efforts have continued to design chiral LC systems [7–11]. Due to the presence of a stereocenter in the molecular unit, mesogens can mainly adopt a helical conformation. In this way, molecular-level chirality is transferred to the macroscopic scale in the LC phase. This leads to exciting applications in advanced displays, sensors, nonlinear optical devices, organic photovoltaic cells, and others [4,12–24]. The most promising LC materials for practical use in display technology possess the SmC_A* phase with a tilt angle close to 45° [25,26]. Such antiferroelectric liquid crystals (AFLCs) have orthoconic properties [27–30] and are described as OAFLCs—orthoconic antiferroelectric liquid crystals. In the surface stabilized geometry (SSOAFLCs), these materials can be considered as an optically negative medium and, at the same time, optically isotropic for a beam of light at normal incidence. This is due to the optical axis of the refractive medium being aligned along the plane of the smectic layer and perpendicular to the boundary surfaces, making it collinear with the light passing through the display at normal incidence. Such SSOAFLC retains all the advantages of AFLC and, most importantly, achieves extremely high optical contrast due to the almost perfectly dark state at zero electric field. This occurs because optical defects remain invisible between crossed polarizers, regardless of any local disorientation of smectic layers that is normal within the SSOAFLC structure. Recent extensive studies have shown a significant influence of the molecular structure of AFLCs on the formation of their orthoconic behavior [31–41]. These properties are typically exhibited by three-ring compounds, often containing fluorine atom(s) as lateral substituents in the benzene ring, with a terminal perfluoroalkoxylalkoxy or perfluoroalkanoyloxyalkoxy chain complemented by various combinations of chirality centers. While (S) enantiomers are most commonly used, several (*R*) enantiomers have recently been synthesized [42,43]. Mixing (*R*) and (*S*) enantiomers results in OAFLC mixtures with a very long helical pitch [44], which is one of the critical parameters for OAFLC applications [45]. The helical pitch for most OAFLCs is shorter than 1.0 μm, and a very thin cell is necessary to obtain a well-oriented smectic layer. Thus, LC materials with a long helical pitch are highly sought.

Here, we present the synthesis of four new (*R*) enantiomers, complemented by a detailed investigation of their chemical and physical properties and potential application in the formulation of mixtures, as they are of high practical importance. This work demonstrates that some synthesized LC materials exhibit near-orthoconic behavior simultaneously with an exceptionally long helical pitch. Further research will undoubtedly focus on their utilization as functional materials in novel multicomponent LC mixtures. So far, only a few mixtures containing (*R*) enantiomers have been formulated. Hence, these materials are undoubtedly a novelty in the group of chiral antiferroelectrics.

2. Materials and Methods

2.1. Method for the Synthesis of (R) Enantiomers

The chemical formula (in general form) for the four esters that were designed and synthesized with (*R*) configuration is presented in Figure 1. These (*R*) enantiomers have the acronym **I.4.(X_1X_2) (*R*)**, where **X_1**, **X_2** = **H** or **F**.

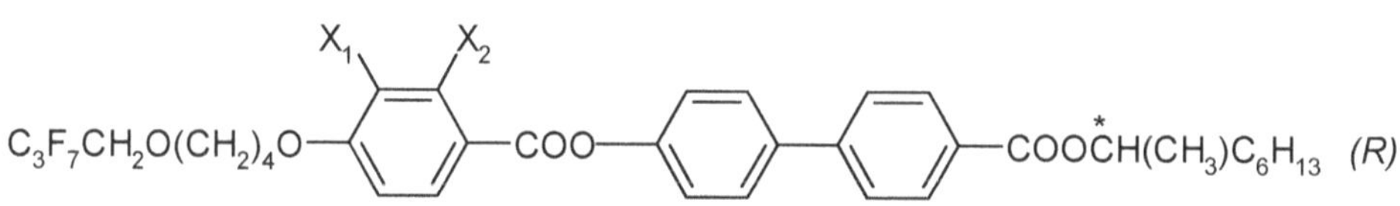

Figure 1. General chemical formula of (*R*) enantiomers; (*) indicates the chiral center.

The enantiomers were synthesized by treating a chiral phenol with benzoic acid chloride in the presence of pyridine, as shown in Figure 2. An efficient preparative method was used to synthesize a chiral hydroxyester of high optical purity [46]. The synthesis was carried out as described in Ref. [47]. The course of the reaction was explained using the example of the enantiomer **I.4.(HH) (*R*)**, which is presented in the Supplementary Materials.

The chemical purity of the synthesized enantiomers was checked first by thin-layer chromatography (there was always one spot on the plate) and then by high-performance liquid chromatography. HPLC-PDA-MS (APCI-ESI dual source) Shimadzu LCMS 2010 EV equipped with a polychromatic UV-VIS detector (Shimadzu Co., Kyoto, Japan) was used to check the purity, and the purity was over 99% for all (*R*) enantiomers. The results of the ionization process of the analyzed enantiomers are presented in Table 1. The MS data and the corresponding purity curves for these enantiomers are shown in Figures S1–S4 (see Supplementary Materials).

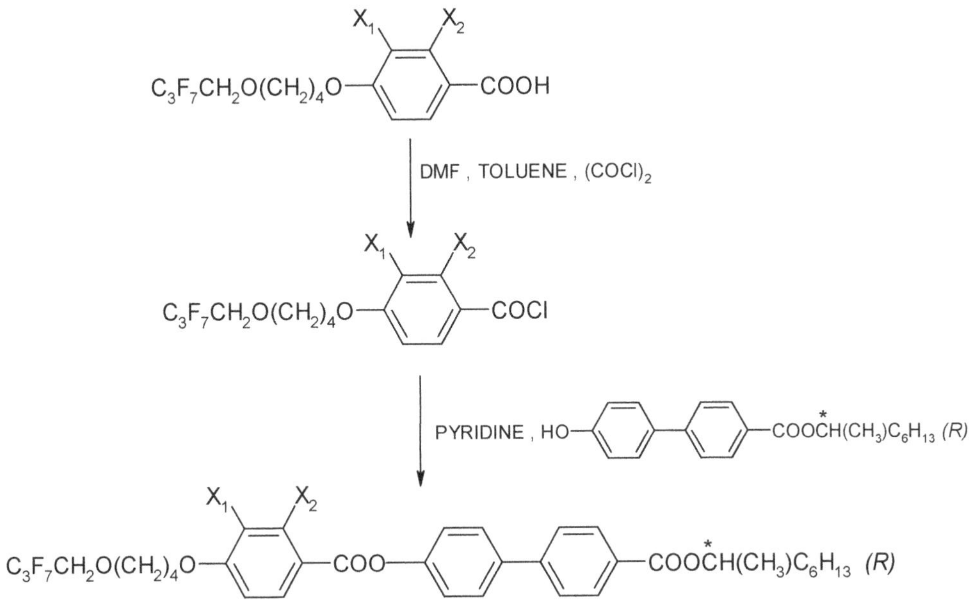

Figure 2. General scheme for the synthesis of (*R*) enantiomers; (*) indicates the chiral center.

Table 1. The acronyms and molecular ions for the synthesized enantiomers.

Acronym of the Enantiomer	Molecular Ion
I.4.(HH) (*R*)	$701[M + H]^-$
I.4.(HF) (*R*)	$718[M - H]^-$
I.4.(FH) (*R*)	$719[M + H]^-$
I.4.(FF) (*R*)	$735[M - H]^-$

2.2. Determination of Optical Purity of (R) Enantiomers

Amylose-based polysaccharide derivatives [48] are the most effective CSPs for enantioseparation of this type of smectic [49,50]. However, in this case, an equally effective immobilized brush-type column with π-electron donor and π-electron acceptor groups was chosen to test the optical purity. Acetonitrile/water 99/1 (v/v) was used as a mobile phase. Isocratic elution was used in the analysis. The enantiomers were dissolved in ACN at 0.6 $mg \cdot mL^{-1}$; the injected sample was 5 μL. HPLC-grade ACN (min. 99.9%) was purchased from a Polish company (POCH S.A., Gliwice, Poland). Double-distilled water was taken (a glass distiller, GFL 2302 model, manufactured in Germany, was used for distillation).

Chiral separation was performed using a Shimadzu LC-20AP HPLC system (Shimadzu Co., Kyoto, Japan). A diode array detector (SPD-M20A) was used. Data were collected using Shimadzu software (Labsolutions, 2010-2017 Shimadzu Co.). The measurements were performed at room temperature (25 °C). The flow rate of the mobile phase was 1.0 $mL \cdot min^{-1}$. The detection wavelength was 254 nm. A ReproSil Chiral NR column with a particle size of 5 μm, surface area of 350 $m^2 \cdot g^{-1}$, dimensions of 250 mm × 4.6 mm i.d, and a pore size of 100 Å (Dr. Maisch, Ammerbuch, Germany) was taken for the enantioseparation. The results of HPLC separation are presented in Figure 3 and Table 2. The resolution (R_s) [51] was higher than 1.5 in each case, indicating a baseline separation.

Table 2. Relative peak areas of (*R*) enantiomers obtained by chiral HPLC and the corresponding enantiomeric excess values of the enantiomers.

Peak Area (%)	I.4.(HH) (*R*)	I.4.(HF) (*R*)	I.4.(FH) (*R*)	I.4.(FF) (*R*)
(*S*)	1.086	0.082	0.081	1.610
(*R*)	98.914	99.918	99.919	98.390
% *ee*	97.83	99.84	99.84	96.78

(**a**)

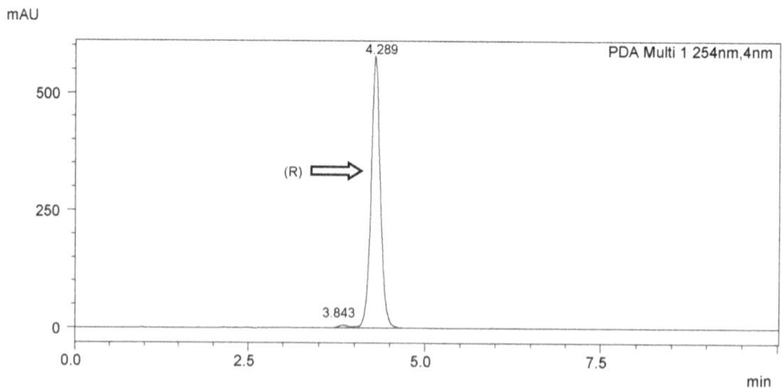

(**b**)

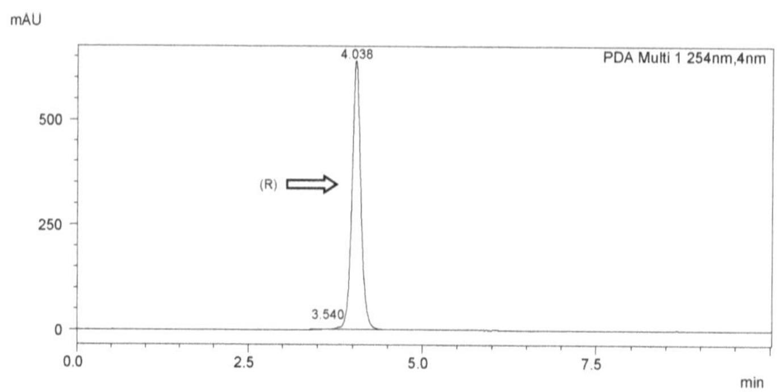

(**c**)

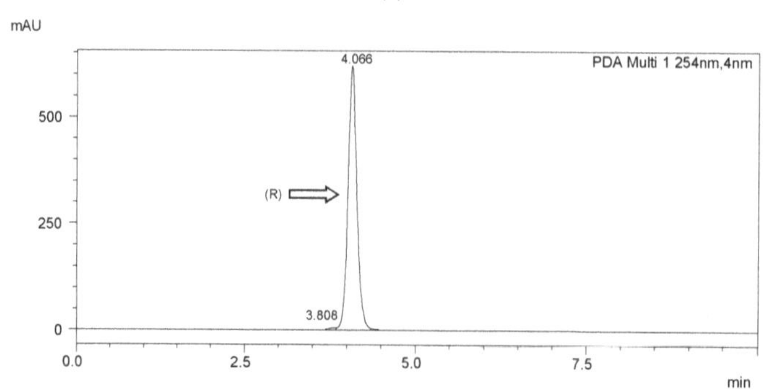

Figure 3. *Cont.*

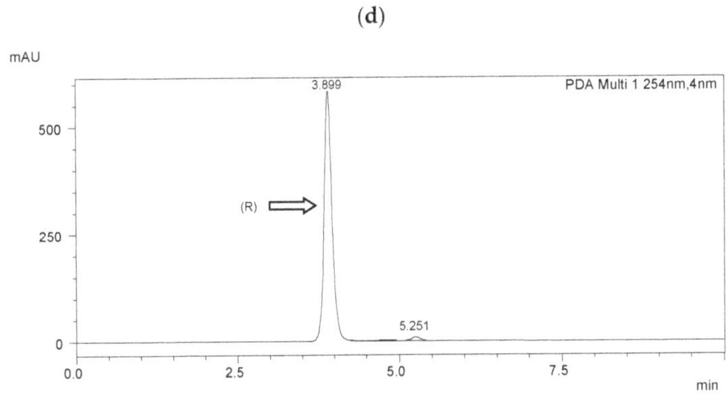

Figure 3. HPLC chromatograms of the enantioseparation of (*R*) enantiomers shown in figures: (a) **I.4.(HH) (*R*)**; (b) **I.4.(HF) (*R*)**; (c) **I.4.(FH) (*R*)**; (d) **I.4.(FF) (*R*)**.

The optical purity of the synthesized (*R*) enantiomers is high, especially for the monofluorinated enantiomers *ee* > 99%. If recrystallized again from anhydrous ethanol, their purity would increase even more, especially with the difluorinated enantiomer. The optical purity of the chiral biphenol used was more than 98%.

2.3. Experimental Methods

The mesophases sequence was determined by polarizing optical microscopy (POM) using an Olympus BX51 polarizing microscope (Shinjuku, Tokyo, Japan) equipped with a Linkam heating/cooling stage THMS-600 (Linkam Scientific Instruments Ltd., Tadworth, UK).

The phase transition temperatures were also precisely determined by differential scanning calorimetry (DSC) using a Perkin-Elmer DSC8000 calorimeter (PerkinElmer, Shelton, CT, USA). The samples of ~3–7 mg, hermetically sealed in aluminum pans, were placed in a nitrogen-filled calorimeter chamber. Temperature and enthalpy change values were calibrated on the extrapolated onset temperatures and the enthalpy changes of the melting points of water, indium, and zinc. The calorimetric measurements were performed on cooling/heating cycles at a rate of 5 $K \cdot min^{-1}$.

The helical pitch [p] was determined on homeotropically aligned samples of materials placed on a single glass plate while leaving the other surface of the sample free. The measurements of the helical pitch were based on the selective light reflection phenomenon [52]. The p(T) characteristics were calculated from the equations $p = \lambda/n$ for the SmC_A* phase and $p = \lambda/2n$ for the SmC* phase (the value of the average refractive index n = 1.5 was estimated according to Ref. [53]). The transmission spectra were acquired using a Shimadzu UV-VIS-NIR spectrometer (wavelength range of 360 nm–3000 nm). AMLWU7 controller with Peltier element was used for temperature control within the 2–110 °C range and accuracy $\pm$ 0.1 K. All measurements were performed in the cooling cycle.

The polarimetry method determined the helical twist sense for two chiral phases [54]. The helical twist sense inversion temperature was determined by analyzing the temperature-dependent transmission of the light through a homeotropically aligned sample under a polarizing optical microscope; the brightest texture indicates an unwound state [34,55].

The tilt angle of the director (Θ) was determined optically using well-aligned planar samples at the bookshelf-like surface stabilized structure by observing the difference between extinction positions at crossed polarizers under opposite d.c. electric fields $\pm$40 $kV \cdot cm^{-1}$. All the measurements were done during the cooling cycle.

The X-ray diffraction method – X'Pert PRO diffractometer (Malvern PANalytical, Malvern, Worcestershire, UK) with a TTK-450 temperature attachment (Anton Paar, Graz, Austria) was used to determine the smectic layer spacing [56]. Before the measurement, the samples were heated until an isotropic liquid was obtained, and then the diffraction patterns were collected in the 2θ = 1.7–30° range (CuKα radiation, Bragg-Brentano geometry) during cooling. Data analysis was performed in WinPLOTR (version from April 2019) [57] and OriginPro 2020b.

3. Results

3.1. Mesomorphic Behavior

In the case of the synthesized liquid crystalline (*R*) enantiomers, the phase sequences were determined based on the characteristic textures and their changes observed in POM. The phase transition temperatures and transition enthalpies were estimated with high accuracy from DSC measurements. The phase sequences and phase transition temperatures, measured during cooling, melting points, m.p., and clearing points, c.p., measured during heating and respective phase transition enthalpies, ΔH, obtained by DSC for (*R*) enantiomers are summarized in Table 3.

Table 3. The phase sequences and phase transition temperatures (in the heating and cooling cycles) [°C] and phase transition enthalpies in brackets [J·g^{-1}] for (*R*) enantiomers determined from DSC in the heating/cooling cycles with a rate of 5 K·min^{-1}. The m.p is bolded, and the c.p. is italicized.

Acronym of the Enantiomer	Cr	T [ΔH]	SmC$_A$*	T [ΔH]	SmC*	T [ΔH]	SmA*	T [ΔH]	Iso
I.4.(HH) (*R*) heating	•	75.4 [+20.9]	•	110.5 [+0.05]	•	*130.6* [+8.5]	–		•
I.4.(HH) (*R*) cooling	•	23.6 [−12.0]	•	104.1 [−0.06]	•	129.1 [−8.1]	–		•
I.4.(HF) (*R*) heating	•	48.7 [+28.8]	•	88.9 [+0.07]	•	*108.5* [+8.3]	–		•
I.4.(HF) (*R*) cooling	•	6.2 [−16.1]	•	83.5 [−0.07]	•	106.6 [−6.7]	–		•
I.4.(FH) (*R*) heating	•	68.4 [+34.6]	•	98.4 [+0.07]	•	115.5 [+0.8]	•	*116.5* [+7.5]	•
I.4.(FH) (*R*) cooling	•	40.6 [−25.9]	•	92.5 [−0.07]	•	114.2 [−0.7]	•	114.8 [−7.7]	•
I.4.(FF) (*R*) heating	•	76.5 [+35.7]	•	98.3 [+0.07]	•	*115.8* [+7.2]	–		•
I.4.(FF) (*R*) cooling	•	56.0 [−33.4]	•	93.9 [−0.06]	•	114.3 [−6.4]	–		•

All four designed enantiomers exhibit the ferroelectric (SmC*) and the antiferroelectric (SmC$_A$*) phases. The paraelectric orthogonal (SmA*) phase is observed in a very short temperature range only for the enantiomer **I.4.(FH) (*R*)**. The polar, synclinic SmC*, and anticlinic SmC$_A$* phases occur over a reasonably broad temperature range. The broadest temperature range of the antiferroelectric phase (above 40 °C) is observed for the monofluorinated enantiomer with **(HF)** type of substitution. The highest clearing point was detected for the unsubstituted enantiomer. The lowest melting and clearing points were detected for enantiomer **I.4.(HF) (*R*)**, see Figure 4. Characteristic microscopic patterns for the SmC$_A$* and SmC* phases are presented in Figure 5a,b. Both the ferroelectric and antiferroelectric phases possess broken fan-shaped textures, and the differences between them are shown for the enantiomer **I.4.(HH) (*R*)**.

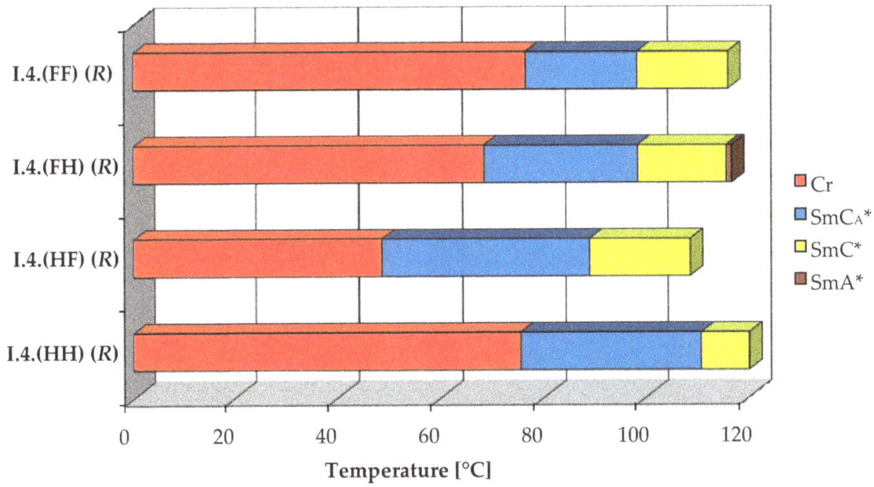

Figure 4. Visualization of phase temperature ranges of (*R*) enantiomers.

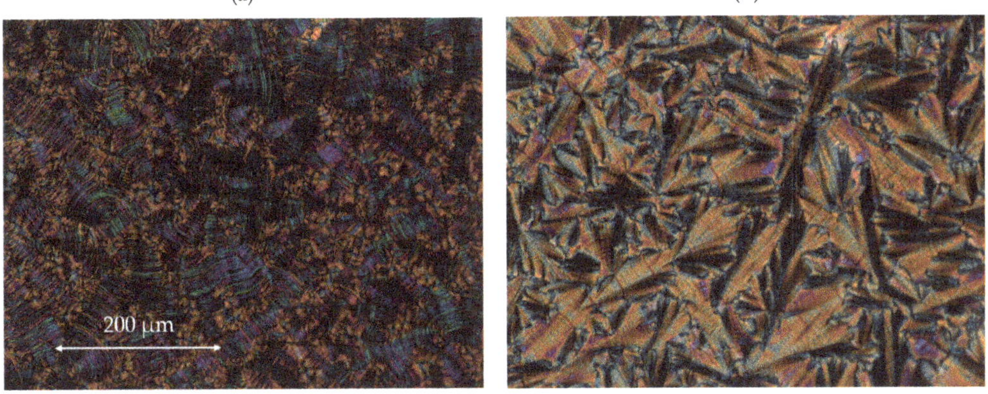

Figure 5. The microphotographs of the microscopic pattern for the enantiomer **I.4.(HH) (*R*)** observed during the cooling cycle: (**a**) in the SmC* phase at 121.1 °C and (**b**) in the SmC_A* phase at 96.7 °C. The width of the microphotographs is ~600 µm.

All the SmC_A*–SmC* transitions exhibit low enthalpy values equal to or below 0.07 $J \cdot g^{-1}$. The SmC* to SmA* phase transition has a higher enthalpy (0.8 $J \cdot g^{-1}$). The highest enthalpy for the clearing point is estimated for the unsubstituted enantiomer—**I.4.(HH) (*R*)**. Melting point enthalpies are pretty low for the compounds **I.4(HF) (*R*)** and **I.4(HH) (*R*)**, while they are significantly higher (greater than 30 $J \cdot g^{-1}$) for the remaining compounds.

The sequence of mesophases was the same when comparing the (*S*) [31] and (*R*) enantiomers. Slightly lower clearing points were observed for (*S*) enantiomers. The source of the differences in phase transition temperatures for (*S*) and (*R*) enantiomers may be these compounds' slightly different enantiomeric purity.

3.2. Helical Pitch and Twist Sense

The behavior of the main parameter of the helical structure, namely, the helical pitch versus temperature, is presented in Figure 6 for (R) enantiomers. Within the ferroelectric SmC* phase, the helical pitch remains almost temperature-independent and does not exceed 300 nm with a left-handed helix. The helix is left-handed in the antiferroelectric phase, but the pitch length increases with temperature. The enantiomer with a fluorine atom in the meta position exhibits a longer helical pitch in all measuring ranges than the unsubstituted enantiomer. For the enantiomers **I.4.(HF) (R)** and **I.4.(FF) (R)**, the helical pitch in the anticlinic SmC_A* phase is out of the measuring range (i.e., above 2.0 μm) of the used spectrophotometer.

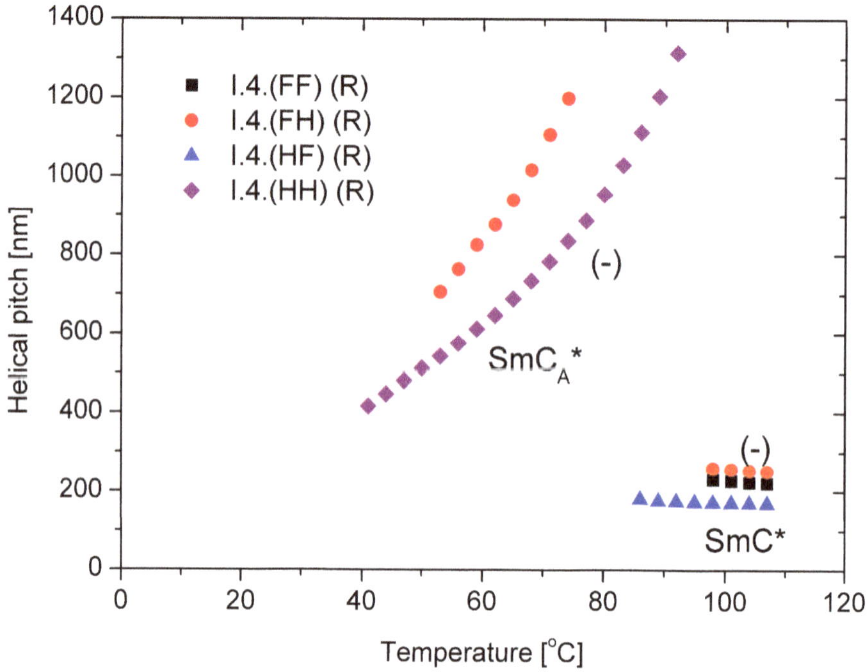

Figure 6. Temperature dependence of the helical pitch length for (*R*) enantiomers; "-" indicates the left-handed helix.

3.3. X-ray Diffraction

The spacing between the smectic layers as a function of temperature for all enantiomers is shown in Figure 7. The layer spacing for the orthogonal SmA* phase was observed only for a single temperature, while in the SmC* and SmC_A* phases, the layer spacing values decreased with decreasing temperature. The layer shrinkage at the SmA*–SmC* transition for the enantiomer **I.4.(FH) (R)** equals ca. 5%. For the enantiomer **I.4.(HH) (R)**, the layer spacing slightly increases just before crystallization, which may be related to the transition to another phase (named hexatic), previously suspected for the enantiomer **I.4.(HH) (S)**.

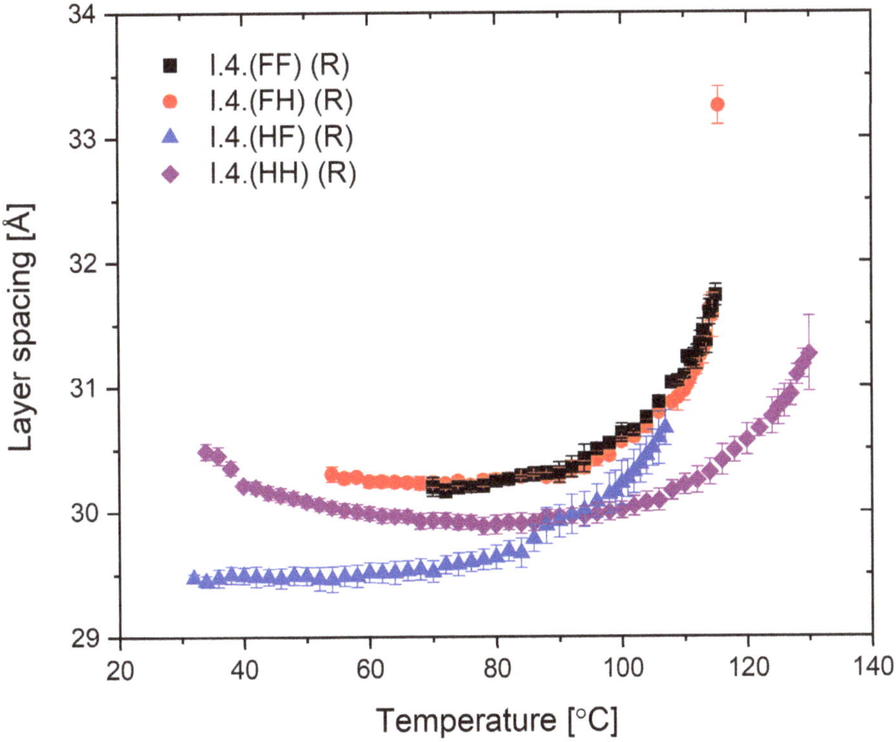

Figure 7. The smectic layer spacing of (R) enantiomers versus temperature.

Most importantly, from an application perspective, layer thickness studies confirm the mesomorphic studies, which show a direct phase transition from the isotropic phase to tilted smectic phases in three of the four enantiomers. This crucial characteristic helps minimize defects in surface-stabilized AFLC geometries [35], improving the optical quality of devices based on such LC materials.

3.4. Tilt Angle

The temperature dependence of the tilt angle (Θ) for all (R) enantiomers is shown in Figure 8.

All enantiomers possess a very high tilt angle of the director in the smectic layers. The highest values of the optical tilt angle (Θ = 43.5° at low temperatures) were found for the monofluorinated enantiomer—**I.4.(HF) (R)**. The lowest Θ values (~40.0°) were found for the difluorinated enantiomer. With increasing temperature, the values of the tilt angle decrease; the highest tilt angle values and their temperature ranges are given in Table 4. Among the studied enantiomers, the unsubstituted enantiomer and the one substituted by a fluorine atom in the ortho position can be considered orthoconics [25,26].

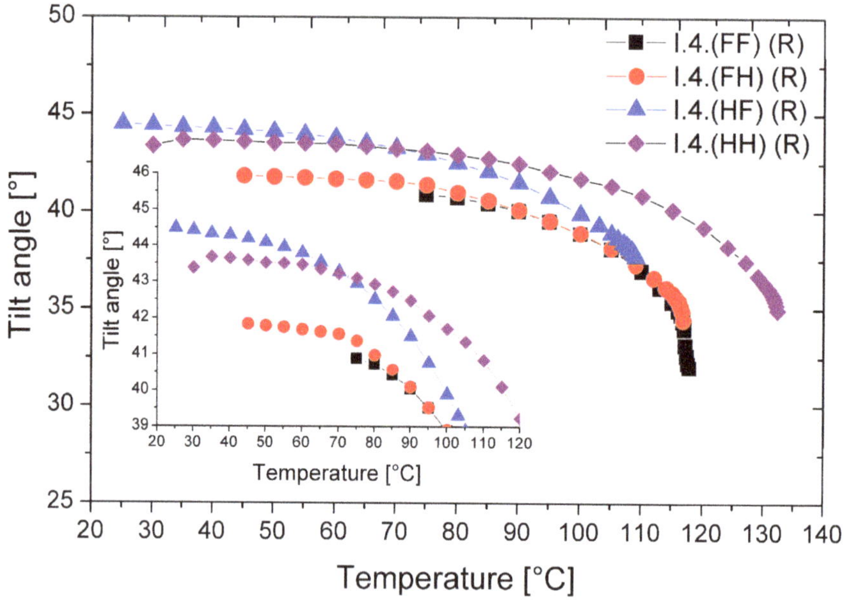

Figure 8. Temperature dependence of the director's tilt angle for (R) enantiomers. The inset enlarges the overlapping curves for better visibility only.

Table 4. The highest tilt angle for (*R*) enantiomers.

Acronym of the Enantiomer	Θ [°]	T [°C]
I.4.(HH) (*R*)	~43.0	30–75
I.4.(HF) (*R*)	~44.0	25–35
I.4.(FH) (*R*)	~41.0	45–75
I.4.(FF) (*R*)	~40.0	75–90

3.5. Properties of Eutectic Mixtures

Two newly synthesized enantiomers and three previously synthesized (*R*) enantiomers [42,43] were used to formulate two antiferroelectric mixtures (denoted as W-446 and W-447); their compositions and phase transition temperatures from DSC measurements obtained from the heating cycle are given in Table 5. The composition of the eutectic mixture was calculated using the equations proposed by Le Chatelier, Schröder, and van Laar, according to Refs. [58–61].

Table 5. The composition and phase transition temperatures [°C] determined on heating for the formulated LC mixtures.

Mixture/Phase Transition Temperatures	Acronym of the Enantiomer	Weight Ratio [%]
W-446/Cr 44.1 SmC_A* 105.4 SmC* 126.0 Iso	**I.4.(HH) (*R*)**	26.34
	I.2.(HH) (*R*)	35.58
	I.6.(HH) (*R*)	38.08
W-447/Cr 27.3 Cr′ 32.9 SmC_A* 91.8 Iso	**I.4.(HF) (*R*)**	35.59
	I.3.(HF) (*R*)	64.41

Both designed mixtures possess the antiferroelectric phase in a wide temperature range. The mixture W-447 directly transitions from the anticlinic to the isotropic phase. Materials with the direct SmC_A*—Iso phase transition usually generate high tilt values rapidly during phase formation; moreover, the layers do not change their spacing too much upon cooling. Hence, the shrinkage of the layer is minimal. Since no phase transitions occur during further cooling, the number of defects is limited, which is essential for applying these materials [60]. Despite the antiferroelectric phase, the mixture W-446 also possesses the ferroelectric phase in a medium temperature range. Both mixtures exhibit relatively low melting points. The clearing point is below 92 °C for the two-component mixture, W-447, and is slightly above 125 °C for the three-component one, W-446.

The helical pitch versus temperature for both mixtures is presented in Figure 9. In the SmC_A* phase, the helical pitch values for both mixtures are comparable at low temperatures. As the temperature increases, the pitch elongates, ultimately reaching complete unwinding at 74 °C for the mixture W-447. The pitch is notably short (i.e., below 200 nm) in the SmC* phase. Additionally, the helix is left-handed in both phases.

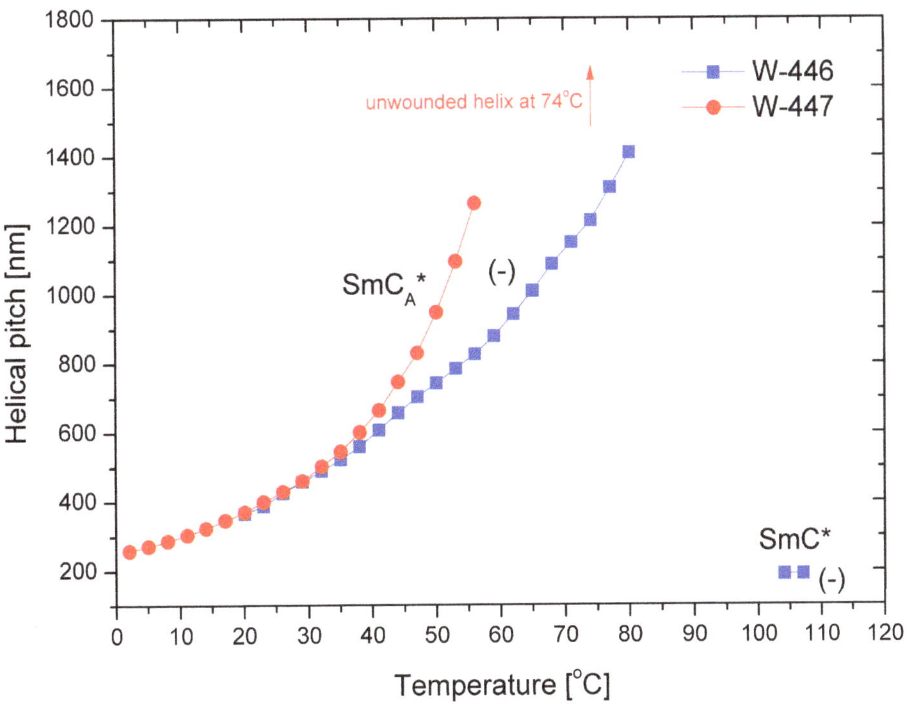

Figure 9. Temperature dependence of the helical pitch length for the mixtures W-446 and W-447; "-" indicates the left-handed helix.

Figure 10 shows the temperature dependencies of the tilt angle of the director for the mixtures W-446 and W-447.

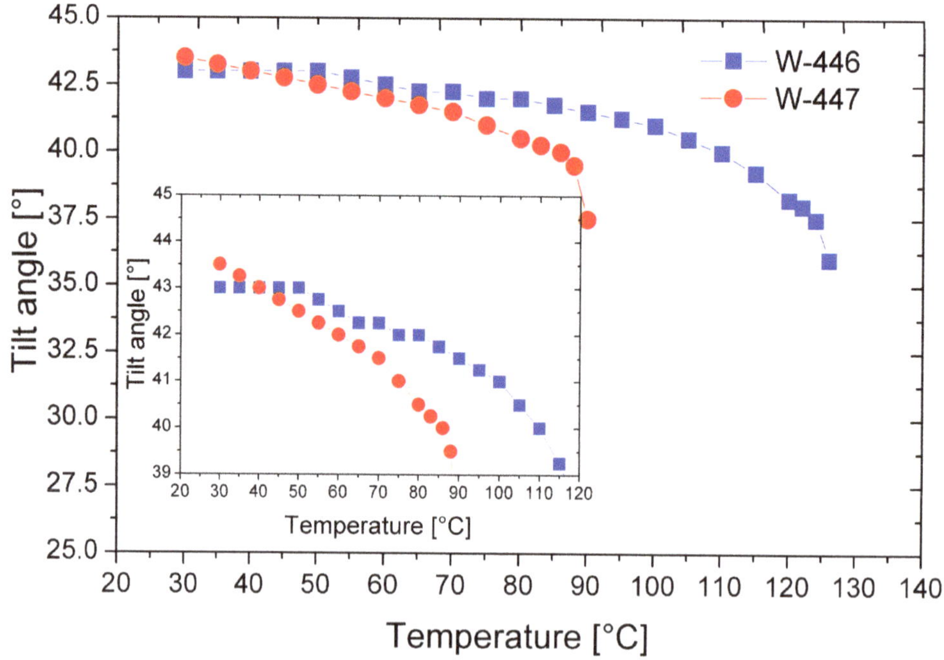

Figure 10. Temperature dependence of the tilt angle of the director for the mixtures W-446 and W-447. The inset enlarges the overlapping curves for better visibility only.

In both mixtures, a high tilt angle of the director is generated just below the phase transition from the isotropic to the smectic phase. As mentioned earlier, this is advantageous, as it minimizes the formation of chevron-type defects in surface-stabilized geometries. Notably, at temperatures below 50 °C, both mixtures exhibit a relatively high tilt angle of the director, exceeding 42 degrees. Furthermore, the mixture W-447 has a tilt angle above 43 degrees near room temperature, significantly enhancing the electro-optic effect's contrast [26].

4. Conclusions

We have synthesized and characterized four liquid crystalline (*R*) enantiomers, each featuring a three-benzene-ring molecular core, a perfluoroalkoxyalkoxy terminal chain, complemented by a chiral center based on (*R*)-(−)-2-octanol. Their mesophase behavior, helical parameters, smectic layer thickness, the tilt angle of the director, and potential for utilization for the LC mixture formulation were systematically studied and evaluated. Two types of phase sequences are observed for the synthesized enantiomers. All enantiomers were found to be enantiotropic LCs and exhibit a broad temperature range of the antiferroelectric phase. The lowest melting points are observed for the monofluorinated esters. In the ferroelectric phase, the helix is left-handed, with almost temperature-independent and very short pitch values. In the antiferroelectric phase for the enantiomers **I.4.(FH) (*R*)** and **I.4.(HH) (*R*)**, the helix is left-handed, and the pitch increases with temperature. In the case of the enantiomers **I.4.(HF) (*R*)** and **I.4.(FF) (*R*)**, the values of helical pitch are above the measuring range of the used spectrophotometer. Most importantly, all the studied enantiomers exhibit the tilt angle of the director, exceeding 40.0° over a broad temperature range. In the case of the enantiomers **I.4.(HH) (*R*)** and **I.4.(HF) (*R*)**, the tilt angle values are sufficiently high to classify them as orthoconic AFLCs.

The antiferroelectric phase characterizes the eutectic mixtures W-446 and W-447, formulated using the designed enantiomers in a very broad temperature range, a long helical pitch, especially at higher temperatures, and they are near-orthoconic AFLCs. Furthermore, the favorable direct SmC_A*−Iso phase transition was detected for one of the designed mixtures.

We can summarize that the synthesized (*R*) enantiomers and previously synthesized (*R*) enantiomers can be very useful for the formulation of LC eutectic mixtures (both binary and multicomponent) for further use in various devices (e.g., in photonics).

Supplementary Materials: The following supporting information can be downloaded at: https://www.mdpi.com/article/10.3390/ma17204967/s1, Figure S1: The mass spectrum and purity curve of the enantiomer denoted as **I.4.(HH) (*R*)**; Figure S2: The mass spectrum and purity curve of the enantiomer denoted as **I.4.(HF) (*R*)**; Figure S3: The mass spectrum and purity curve of the enantiomer denoted as **I.4.(FH) (*R*)**; Figure S4: The mass spectrum and purity curve of the enantiomer denoted as **I.4.(FF) (*R*)**.

Author Contributions: Conceptualization, M.U.; methodology, M.U.; software, M.U.; M.Z., M.C., P.M., A.B. and A.D.; validation, M.U.; formal analysis, M.U.; investigation, M.U., M.Z., M.C., P.M., A.B. and A.D.; resources, M.U.; data curation, M.U.; writing—original draft preparation, M.U., M.C., A.B. and A.D.; writing—review and editing, M.U.; visualization, M.U.; M.Z., M.C., A.B. and A.D.; supervision, M.U. All authors have read and agreed to the published version of the manuscript.

Funding: The study was funded by a University Research Grant [No. 22-720]. A.B. also acknowledges the financial support (DSC measurements of chiral LCs) from the Czech Science Foundation [Project No. CSF 23-42944L].

Institutional Review Board Statement: Not applicable.

Informed Consent Statement: Not applicable.

Data Availability Statement: The original contributions presented in the study are included in the article and Supplementary Materials, further inquiries can be directed to the corresponding author.

Conflicts of Interest: The authors declare no conflicts of interest.

References

1. Demus, D.; Goodby, J.; Gray, G.W.; Spiess, H.-W.; Vill, H.-W.; Lagerwall, S.T. *Handbook of Liquid Crystals*; Wiley-VCH: Weinheim, Germany, 1998; Volume 2B, Chapter VI 2, pp. 515–664.
2. Collings, P.J. *Liquid Crystals: Nature's Delicate Phase of Matter*; Princeton University Press: Princeton, NJ, USA, 1990.
3. Goodby, J.W.; Saez, I.M.; Cowling, S.J.; Görtz, V.; Draper, M.; Hall, A.W.; Sia, S.; Cosquer, G.; Lee, S.-E.; Raynes, E.P. Transmission and amplification of information and properties in nanostructured liquid crystals. *Angew. Chem. Int. Ed. Engl.* **2008**, *47*, 2754–2787. [CrossRef] [PubMed]
4. Lagerwall, J.P.F.; Giesselmann, F. Current Topics in Smectic Liquid Crystal Research. *ChemPhysChem* **2006**, *7*, 20–45. [CrossRef] [PubMed]
5. Meyer, R.B.; Liebert, L.; Strzelecki, L.; Keller, P. Ferroelectric liquid crystals. *Le J. Phys. Lett.* **1975**, *36*, 69–71. [CrossRef]
6. Chandani, A.D.L.; Górecka, E.; Ouchi, Y.; Takezoe, H.; Fukuda, A. Antiferroelectric chiral smectic phases responsible for the tristable switching in MHPOBC. *Jpn. J. Appl. Phys.* **1989**, *28*, L1265–L1268. [CrossRef]
7. Lemieux, R.P. Molecular recognition in chiral smectic liquid crystals: The effect of core–core interactions and chirality transfer on polar order. *Chem. Soc. Rev.* **2007**, *36*, 2033–2045. [CrossRef]
8. Goodby, J.W.; Chin, E. A chiral induced ferroelectric liquid crystal phase transition with a vanishingly small enthalpy. *Liq. Cryst.* **1988**, *3*, 1245–1254. [CrossRef]
9. Goodby, J.W.; Waugh, M.A.; Stein, S.M.; Chin, E.; Pindak, R.; Patel, J.S. Characterization of a new helical smectic liquid crystal. *Nature* **1989**, *337*, 449–452. [CrossRef]
10. Pramanik, A.; Das, M.K.; Das, B.; Dąbrowski, R. Preparation and Study of the Electro-optical Properties of Binary Mixtures of Orthoconic Antiferroelectric Esters and Achiral Phenyl Pyrimidine Liquid Crystal. *Soft Mater.* **2015**, *13*, 201–209. [CrossRef]
11. Nowicka, K.; Dardas, D.; Garbat, K. High permittivity of chiral tilted smectic phases in a binary mixture. *Phase Trans.* **2013**, *86*, 153–160. [CrossRef]
12. Ha, N.Y.; Jeong, S.M.; Nishimura, S.; Suzaki, G.; Takanishi, Y.; Ishikawa, K.; Takezoe, H. Fabrication of a simultaneous red–green–blue reflector using single-pitched cholesteric liquid crystals. *Nat. Mater.* **2008**, *7*, 43–47. [CrossRef]
13. Busson, P.; Örtegren, J.; Ihre, H.; Gedde, U.W.; Hult, A.; Andersson, G.; Eriksson, A.; Lindgren, M. Preparation of Mesogen-Functionalized Dendrimers for Second-Order Nonlinear Optics. *Macromolecules* **2002**, *35*, 1663–1671. [CrossRef]

14. Yamane, S.; Sagara, Y.; Kato, T. A thermoresponsive photoluminescent smectic liquid crystal: Change of photoluminescent color on the smectic–smectic phase transition. *Chem. Commun.* **2009**, *24*, 3597–3599. [CrossRef] [PubMed]
15. Walba, D.M.; Dyer, D.J.; Sierra, T.; Cobben, P.L.; Shao, R.; Clark, N.A. Ferroelectric Liquid Crystals for Nonlinear Optics: Orientation of the Disperse Red 1 Chromophore along the Ferroelectric Liquid Crystal Polar Axis. *J. Am. Soc.* **1996**, *118*, 1211–1212. [CrossRef]
16. Nowicka, K.; Kuczyński, W. Non-linear electro-optical effects in the study of the helical smectic liquid crystals. *Phase Trans.* **2016**, *89*, 376–382. [CrossRef]
17. Lembrikov, B.I.; Ianetz, D.; Ben-Ezra, Y. Nonlinear Optical Phenomena in a Silicon-Smectic A Liquid Crystal (SALC) Waveguide. *Materials* **2019**, *12*, 2086. [CrossRef] [PubMed]
18. Yoshizawa, A. Ferroelectric Smectic Liquid Crystals. *Crystals* **2024**, *14*, 350. [CrossRef]
19. Balasubramanian, V.; Vijayakumar, V.N.; Vasanthi, T. Photonic applications of linear double hydrogen bond ferroelectric liquid crystal mixture: Optical modulators. *Braz. J. Phys.* **2023**, *53*, 36. [CrossRef]
20. Srivastava, A.K.; Chigrinov, V.G.; Kwok, H.S. Ferroelectric liquid crystals: Excellent tool for modern displays and photonics. *J. SID* **2015**, *23*, 233–283. [CrossRef]
21. Yamamoto, N.; Koshoubu, N.; Mori, K.; Nakamura, K.; Yamada, Y. Full-colour antiferroelectric liquid crystal display. *Ferroelectrics* **1993**, *149*, 295. [CrossRef]
22. Lagerwall, J.P.F.; Scalia, G. A new era for liquid crystal research: Applications of liquid crystals in soft matter nano-, bio- and microtechnology. *Curr. Appl. Phys.* **2012**, *12*, 1387–1412. [CrossRef]
23. Chen, S.; Chen, Y.; Tong, X.; Wu, B.; Ma, M.; Shi, Y.; Wang, X. Room temperature optical image storage devices based on novel photo-responsive chiral azobenzene liquid crystal dopants. *Mater. Res. Express* **2016**, *3*, 115701. [CrossRef]
24. Bubnov, A.; Iwan, A.; Cigl, M.; Boharewicz, B.; Tazbir, I.; Wójcik, K.; Sikora, A.; Hamplová, V. Photosensitive self-assembling materials as functional dopants for organic photovoltaic cells. *RSC Adv.* **2016**, *6*, 11577–11590. [CrossRef]
25. D'have, K.; Dahlgren, A.; Rudquist, P.; Lagerwall, J.P.F.; Anderson, G.; Matuszczyk, M.; Lagerwall, S.T.; Dąbrowski, R.; Drzewiński, W. Antiferroelectric liquid crystals with 45° tilt a new class of promising electro-optic materials. *Ferroelectrics* **2000**, *244*, 115–128. [CrossRef]
26. Rudquist, P. Orthoconic antiferroelectric liquid crystals. *Liq. Cryst.* **2013**, *40*, 1678–1697. [CrossRef]
27. D'have, K.; Rudquist, P.; Lagerwall, S.T.; Pauwels, H.; Drzewiński, W.; Dąbrowski, R. Solution of the dark state problem in antiferroelectric liquid crystal displays. *App. Phys. Lett.* **2000**, *76*, 3528–3530. [CrossRef]
28. Castillo, P.L.; Otón, J.M.; Dąbrowski, R.; Lara, A.; Quintana, X.; Bennis, N. Electrooptics of antiferroelectric orthoconic reflective displays. *Proc. SPIE* **2004**, *5565*, 284–289.
29. Piecek, W.; Bubnov, A.; Perkowski, P.; Morawiak, P.; Ogrodnik, K.; Rejmer, W.; Żurowska, M.; Hamplová, V.; Kašpar, M. An effect of structurally non compatible additive on the properties of a long pitch antiferroelectric orthoconic mixture. *Phase Trans.* **2010**, *83*, 551–563. [CrossRef]
30. Agrahari, K.; Nautiyal, V.K.; Vimal, T.; Pandey, S.; Kumar, S.; Manohar, R. Modification in different physical parameters of orthoconic antiferroelectric liquid crystal mixture via the dispersion of hexanethiol capped silver nanoparticles. *J. Mol. Liq.* **2021**, *332*, 115840. [CrossRef]
31. Żurowska, M.; Dąbrowski, R.; Dziaduszek, J.; Garbat, K.; Filipowicz, M.; Tykarska, M.; Rejmer, W.; Czupryński, K.; Spadło, A.; Bennis, N.; et al. Influence of alkoxy chain length and fluorosubstitution on mesogenic and spectral properties of high tilted antiferroelectric esters. *J. Mater. Chem.* **2011**, *21*, 2144–2153. [CrossRef]
32. Dąbrowski, R.; Gąsowska, J.; Otón, J.M.; Piecek, W.; Przedmojski, J.; Tykarska, M. High tilted antiferroelectric liquid crystalline materials. *Displays* **2004**, *25*, 9–19. [CrossRef]
33. Dąbrowski, R.; Kula, P.; Raszewski, Z.; Piecek, W.; Otón, J.M.; Spadło, A. New orthoconic antiferroelectrics useful for applications. *Ferroelectrics* **2010**, *395*, 116–132. [CrossRef]
34. Czerwiński, M.; Tykarska, M. Helix parameters in bi- and multicomponent mixtures composed of orthoconic antiferroelectric liquid crystals with three ring molecular core. *Liq. Cryst.* **2014**, *41*, 850–860. [CrossRef]
35. Milewska, K.; Drzewiński, W.; Czerwiński, M.; Dąbrowski, R.; Piecek, W. Highly tilted liquid crystalline materials possessing a direct phase transition from antiferroelectric to isotropic phase. *Mater. Chem. Phys.* **2016**, *171*, 33–38. [CrossRef]
36. Garg, B.; Shariq, M.; Alathlawi, H.J.; Almutib, E.; Alshareef, T.H.; Alzahrani, A.; Khan, M.S.; Slimani, Y. Comparative computational analysis of orthoconic antiferroelectric liquid crystals: DFT analysis. *J. Mol. Mod.* **2024**, *30*, 328. [CrossRef]
37. Dwivedi, A.; Dwivedi, S.; Pandey, M.B.; Dąbrowski, R.; Dhar, R. Collective molecular relaxations and electro-optical switching response of a wide room temperature antiferroelectric liquid crystal mixture. *Opt. Mater.* **2024**, *149*, 115016. [CrossRef]
38. Gupta, D.; Kula, P.; Bhattacharjee, A. Mesomorphic, electro-optic and dielectric behaviour of a semi-fluorinated chiral liquid crystalline material forming polar smectic phases. *J. Mol. Struct.* **2020**, *1219*, 128557. [CrossRef]
39. Lagerwall, S.T. Orthoconic liquid crystals—A case study. *Adv. Coll. Interface Sci.* **2014**, *208*, 1–9. [CrossRef] [PubMed]
40. Piecek, W. Molecular Structures and the Induction of the Orthoconic Properties in Antiferroelectric Liquid Crystals. *Ferroelectrics* **2010**, *395*, 108–115. [CrossRef]
41. Piecek, W.; Kula, P.; Raszewski, Z.; Perkowski, P.; Morawiak, P.; Kędzierski, J.; Dąbrowski, R.; Sun, X.W. An Influence of a Single Fluorine Atom Position in the Molecular Rigid Core on Physical Properties of Orthoconic Antiferroelectric Liquid Crystal. *Ferroelectrics* **2008**, *365*, 78–87. [CrossRef]

42. Żurowska, M.; Dziaduszek, J.; Szala, M.; Morawiak, P.; Bubnov, A. Effect of lateral fluorine substitution far from the chiral center on mesomorphic behaviour of highly titled antiferroelectric (S) and (R) enantiomers. *J. Mol. Liq.* **2018**, *267*, 504–510. [CrossRef]
43. Urbańska, M.; Strzeżysz, O.; Szala, M. Highly tilted antiferroelectric (R) enantiomers useful for the formulation of eutectic mixtures. *Liq. Cryst.* **2020**, *47*, 179–190. [CrossRef]
44. Urbańska, M.; Morawiak, P.; Czerwiński, M. Effect of doping by enantiomers with the different absolute configuration and phase sequence on mesomorphic, helical and electro-optical properties of highly tilted chiral anticlinic mixture. *J. Mol. Liq.* **2020**, *309*, 113141. [CrossRef]
45. Tykarska, M.; Stolarz, Z.; Dziaduszek, J. Studies of helical pitch in antiferroelectric compounds and mixtures. *Ferroelectrics* **2004**, *311*, 51–57. [CrossRef]
46. Drzewiński, W.; Dąbrowski, R.; Czupryński, K. Orthoconic antiferroelectrics. Synthesis and mesomorphic properties of optically active (S)-(+)-4-(1-methylheptyloxycarbonyl)phenyl 4′-(fluoroalkanoyloxyalkoxy)biphenyl-4-carboxylates and 4′-(alkanoyloxyalkoxy)biphenyl-4-carboxylates. *Polish J. Chem.* **2002**, *76*, 273–284. [CrossRef]
47. Żurowska, M.; Dąbrowski, R.; Dziaduszek, J.; Czupryński, K.; Skrzypek, K.; Filipowicz, M. Synthesis and mesomorphic properties of chiral esters comprising partially fluorinated alkoxyalkoxy terminal chains and a 1-methylheptyl chiral moiety. *Mol. Cryst. Liq. Cryst.* **2008**, *495*, 145–157. [CrossRef]
48. Teixeira, J.; Tiritan, M.E.; Pinto, M.M.M.; Fernandes, C. Chiral Stationary Phases for Liquid Chromatography: Recent Developments. *Molecules* **2019**, *24*, 865. [CrossRef]
49. Vaňkátová, P.; Kubíčková, A.; Kalíková, K. Enantioseparation of liquid crystals and their utilization as enantiodiscrimination materials. *J. Chromatogr. A* **2022**, *1673*, 463074. [CrossRef]
50. Vaňkátová, P.; Kalíková, K.; Kubíčková, A. Advantages of polar organic solvent chromatography for enantioseparation of chiral liquid crystals. *J. Chromatogr. A* **2023**, *1709*, 464383. [CrossRef] [PubMed]
51. Ravisankar, P.; Anusha, S.; Supriya, K.; Kumar, U.A. Fundamental Chromatographic Parameters. *Inter. J. Pharm. Sci. Rev. Res.* **2019**, *55*, 46–50.
52. Takezoe, H.; Kondo, K.; Fukuda, A.; Kuze, E. Determination of helical pitch in homeotropic cell of chiral Smectic C liquid crystal using center laser. *Jpn. J. Appl. Phys.* **1982**, *21*, L627–L629. [CrossRef]
53. Raszewski, Z.; Kędzierski, J.; Perkowski, P.; Piecek, W.; Rutkowska, J.; Kłosowicz, S.; Zieliński, J. Refractive indices of the MHPB(H)PBC and MHPB(F)PBC antiferroelectric liquid crystals. *Ferroelectrics* **2002**, *276*, 289–300. [CrossRef]
54. Kuczyński, W.; Lagerwall, S.T.; Matuszczyk, M.; Skarp, K.; Stebler, B.; Wahl, J. Fast switching low-temperature liquid crystal mixture. *Mol. Cryst. Liq. Cryst.* **1987**, *146*, 173–187. [CrossRef]
55. Tykarska, M.; Czerwiński, M. The inversion phenomenon of helical twist sense in antiferroelectric liquid crystals phase from electronic and vibrational circular dichroism. *Liq. Cryst.* **2016**, *46*, 462–472. [CrossRef]
56. Vertogen, G.; de Jeu, W.H. *Thermotropic Liquid Crystals, Fundamentals*; Springer: Berlin/Heidelberg, Germany, 1988.
57. Roisnel, T.; Rodriguez-Carvajal, J. WinPLOTR: A Windows Tool for Powder Diffraction Pattern Analysis. *Mater. Sci. Forum* **2001**, *378–381*, 118–123. [CrossRef]
58. Schröder, I.Z. Über die Abhängigkeit der Löslichkeit eines festen Körpers von seiner Schmelztemperatur. *Z. Für Phys Chem.* **1893**, *11*, 449–465. [CrossRef]
59. van Laar, J.J. Die Schmelz- oder Erstarrungskurven bei binären Systemen, wenn die feste Phase ein Gemisch (amorphe feste Lösung oder Mischkristalle) der beiden Komponenten ist. *Z. Für Phys Chem.* **1908**, *63*, 257–297. [CrossRef]
60. Le Chatelier, H. Sur un énoncé générale des lois des équlibres chimiques. *C. R. Acad. Sci.* **1884**, *99*, 786–789.
61. de Matos, U.P.R.; Lopes, L.D.; Scheeren, A.P.; Schneider, J.M.F.M.; Bortoluzzi, A.J.; Bonacorso, H.G.; Gonçalves, I.L.; Merlo, A.A. Modeling of binary mixtures by applying the Schröeder-van Laar equation to a set of substances with structural diversity. *Quim. Nov.* **2024**, *48*, 1–13.

materials

MDPI

Article

Strong Impact of Particle Size Polydispersity on the Thermal Conductivity of Yukawa Crystals

Konstantin V. Tretiakov [1,2,*] and Krzysztof Hyżorek [3]

[1] Institute of Molecular Physics, Polish Academy of Sciences, M. Smoluchowskiego 17, 60-179 Poznań, Poland
[2] Faculty of Technology, University of Kalisz, Nowy Świat 4, 62-800 Kalisz, Poland
[3] Game Physics Solutions—Krzysztof Hyżorek, Osiniec 33i, 62-200 Osiniec, Poland; krzysztof.hyzorek@gmail.com
* Correspondence: tretiakov@ifmpan.poznan.pl

Abstract: Control of thermal transport in colloidal crystals plays an important role in modern technologies. A deeper understanding of the governing heat transport processes in various systems, such as polydisperse colloidal crystals, is required. This study shows how strongly the particle size polydispersity of a model colloidal crystal influences the thermal conductivity. The thermal conductivity of model colloidal crystals has been calculated using molecular dynamics simulations. The model crystals created by particles interacting through Yukawa (screened-Coulomb) interaction are assumed to have a face-centered cubic structure. The influence of the Debye screening length, contact potential, and particle size polydispersity on the thermal conductivity of Yukawa crystals was investigated. It was found that an increase in particle size polydispersity causes a strong—almost fivefold—decrease in the thermal conductivity of Yukawa crystals. In addition, the obtained results showed that the effect of the particle size polydispersity on reducing the thermal conductivity of Yukawa crystals is stronger than changes in values of the Debye screening length or the contact potential.

Keywords: thermal conductivity; colloid crystal; Yukawa potential; particle size polydispersity; molecular dynamics simulations

1. Introduction

Interest in colloidal crystals [1] is increasing in light of their possible use in optoelectronics [2], photonics [3], and medicine [4]. To obtain materials with the desired physical properties, research is carried out on colloids composed of spherical and complex molecules, such as "dumbbells" [5], and mixtures of spherical and cylindrical molecules [6]. Electric or magnetic fields are used to control the self-assembly of colloidal crystals [7,8] and to obtain the desired structures by changing the orientation of molecules [8,9]. In condensed-matter physics, the Yukawa potential is widely used to describe the interaction between particles [10]. The screened Coulomb "Yukawa" interaction [11,12] is quite successfully used to describe suspensions of charge-stabilized colloids as long as the standard Derjaguin–Landau–Verwey–Overbeek theory [13] is applicable and the van der Waals interactions are negligible. This repulsive screened-Coulomb potential is a special case of the classical effective pair interaction potential of the single-component colloid model, which describes the interaction between an isolated pair of spherical particles of finite diameter. The particles, which interact through the Yukawa potential with a hard core, form liquid phase and solid phase with body-centered cubic (bcc) or face-centered cubic (fcc) structures depending on the Debye screening length and volume fraction. All those phases were predicted by theory [14,15] and observed both experimentally [16] and in simulations [17,18]. Recently, it has been also shown that Yukawa crystals with a face-centered cubic structure exhibit auxetic properties [19,20].

One of the characteristic features of colloidal systems is the particle size polydispersity [21,22] associated with the synthesis of colloidal particles. That polydispersity deeply affects various thermodynamic properties. An increase in the particle size polydispersity significantly affects the phase diagram of Yukawa systems by shifting the phase transition toward higher packing fractions [23,24]. The influence of polydispersity on the elastic properties of systems [25], in which particles interact through hard [26] and soft (inverse power) [27] interaction potentials, has also been investigated. The influence of size polydispersity on the auxetic properties of Yukawa crystals turned out to be particularly interesting, as it was shown that an increase in size polydispersity leads to the strengthening of the auxetic properties of the system [28]. Further research on the influence of the polydispersity of particle sizes on various physical properties is desirable and may contribute to their possible practical application.

This research is devoted to studying heat transport in a model colloidal system. The thermal conductivity of materials plays an important role in modern technologies. Materials with high thermal conductivity are essential for applications in electronic devices because of their miniaturization, which increases the local power density and leads to temperature growth [29,30]. On the other hand, materials with low thermal conductivity are used in many areas such as clothing, refrigeration, insulation, and thermoelectrics [31]. In bulk materials, high thermal conductivity is mostly observed in crystalline solids with well-defined crystal lattices in which phonons are responsible for heat transport [29,30], whereas disordered and amorphous materials usually have low thermal conductivities. In such materials, the phonon's mean free path is reduced by increased scattering at the disordered structure and leads to a decrease in thermal transport [32]. In addition, the boundary and interface scattering also affect the thermal conductivity in organic [33] and inorganic [34,35] materials. Regarding colloidal crystals, the geometrical constrictions at the interparticle contact points mainly govern their thermal conductivity, and also the thermal properties of the materials of colloidal particles [36,37]. Most colloidal assembly structures are periodically ordered materials; however, the controlled introduction of the disorder can lead to changes in their physical properties; for instance, in phononic bandgaps in the hypersonic regime [38]. Recently, the influence of structural order on the thermal transport properties of binary mixtures of monodisperse particles with varying size ratios was investigated. It has been shown that disorder in a binary colloidal glass can reduce the thermal conductivity by about 50% compared to the crystalline phase [39]. The question arises as to how the thermal conductivity of a colloidal crystal changes without glass transition while maintaining a regular crystal structure, for example, fcc, but only due to the polydispersity of particle sizes. In other words, how will a local disorder affect the thermal conductivity as opposed to the previously studied effect of global disorder in a binary colloidal glass? A deeper understanding of the impact of local disorder on thermal transport at colloids is important from a practical and theoretical point of view, and Yukawa crystals seem to be an appropriate model system for this purpose. To the best of our knowledge, the effect of varying the size of colloid particles on the thermal conductivity of the polydisperse colloidal crystal has not been investigated so far. This contribution closes this gap and shows how the particle size polydispersity influences the thermal conductivity of model colloidal crystal.

This work's main aim is to determine the thermal conductivity of the face-centered cubic Yukawa crystals at various Debye screening lengths. In particular, an important aspect of this research is to estimate the effect of particle size polydispersity on the thermal conductivity of Yukawa crystals.

This article is organized as follows. In Section 2, the studied model, the research method, and simulation details are described. Section 3 is devoted to a discussion of the obtained results, whereas Section 4 contains the conclusions.

2. Model and Method

2.1. Model

We studied a model system which consists of N particles that form a face-centered cubic lattice under periodic boundary conditions. An interaction between colloidal particles can be accurately described by models in which particles interact through the hard-core repulsive Yukawa potential (HCRYP) [16–18]. This potential combines the hard-core part of the potential responsible for the interaction between the "cores" of colloidal particles associated with their finite size and the Yukawa potential which describes screened-Coulomb repulsion between the charged colloidal particles. In reality, the sterically stabilized, polymethyl methacrylate colloids [16] can be successfully described by Yukawa's potential and mimicked by Yukawa crystals.

Original HCRYP [16–18] is highly nonharmonic because it contains hard interactions. In molecular dynamics (MDs) simulations, considering the hard interaction requires special treatment [40] that unnecessarily complicates the simulation. Determination of the thermal conductivity coefficient requires long simulation runs to achieve high calculation statistics, and using such a procedure would make the research more difficult. Therefore, part of the potential responsible for the finite size of the particle (hard-core potential) was replaced with an inverse power-law potential with exponent $n = 96$, which was successfully used to describe hard walls [35,41] when determining the thermal conductivity of systems with argon. In the present study, we use an approximation of the HCRYP of the following form

$$\beta u(r_{ij}) = \beta\varepsilon\frac{\sigma_i\sigma_j}{\langle\sigma\rangle r_{ij}}e^{-\kappa(r_{ij}-\sigma')} + \beta\varepsilon\left(\frac{\sigma'}{r_{ij}}\right)^n \tag{1}$$

where σ_i and σ_j are the diameters of interacting particles, $\sigma' = (\sigma_i + \sigma_j)/2$, and $\langle\sigma\rangle$ is the mean particle diameter in the system, $\beta = 1/(k_B T)$, T is the temperature, k_B is the Boltzmann constant, ε is the contact potential, and k^{-1} is the Debye screening length. We consider systems in which particles exhibit dispersion of size that is described by the parameter of polydispersity. The polydispersity is defined as the standard deviation of the particle size distribution divided by the mean [21,23]

$$\delta = \frac{\sqrt{\langle\sigma^2\rangle - \langle\sigma\rangle^2}}{\langle\sigma\rangle}. \tag{2}$$

In real colloidal systems, the distribution of particle diameters can be approximated by Gaussian or log-normal distribution [21,42]. In our work, we took the Gaussian distribution of the diameters of particles as in previous works [21,26,28]. It should be pointed out that for $\delta = 0$, we have a system of particles with identical diameters that is further referred to as a monodisperse Yukawa system [17–20]. The models of face-centered cubic structures of monodisperse and polydisperse Yukawa crystals are shown in Figure 1.

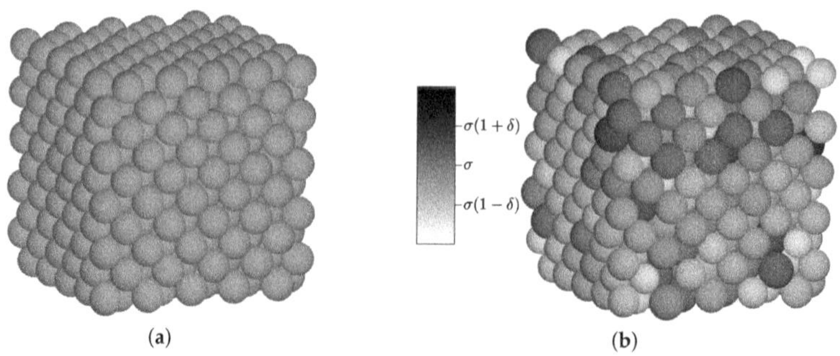

Figure 1. The face-centred cubic Yukawa crystals: (**a**) monodisperse (**b**) polydisperse (with $\delta = 8\%$).

The polydisperse fcc Yukawa crystals with a given polydispersity parameter (δ) were obtained based on structures fulfilling the following conditions

$$|\langle\sigma\rangle - 1| \leq 10^{-5}, \tag{3}$$

$$\left|\frac{\sqrt{\langle\sigma^2\rangle - \langle\sigma\rangle^2}}{\langle\sigma\rangle} - \delta\right| \leq 10^{-6}. \tag{4}$$

2.2. Method

We determined the thermal conductivity of Yukawa crystals using equilibrium molecular dynamics (EMDs) simulations according to the Green–Kubo formula [43–45]. The general form of the Green–Kubo relation, according to the linear response theory [40,46], can be written as follows [40,43,44]

$$K = \int_0^\infty \langle \dot{A}(t)\dot{A}(0)\rangle dt, \tag{5}$$

where K is a transport coefficient and $\dot{A}$ is time derivative of a corresponding microscopic variable. The Green–Kubo formula for the thermal conductivity reads as [46]

$$\lambda = \frac{\beta}{VT}\int_0^\infty \langle \mathbf{j}(0)\mathbf{j}(t)\rangle dt, \tag{6}$$

where V is the volume, the angular brackets denote the average over time in the case of a EMDs simulation, and $\mathbf{j}$ is the microscopic heat current given as [46,47]

$$\mathbf{j}(t) = \sum_i \mathbf{v}_i\epsilon_i + \frac{1}{2}\sum_i \sum_{j,i\neq j} \mathbf{r}_{ij}(\mathbf{F}_{ij}\cdot\mathbf{v}_i), \tag{7}$$

where $\mathbf{r}_{ij}$ the distance between particles i and j, $\mathbf{v}_i$ is the velocity of particle i, and $\mathbf{F}_{ij}$ is the force exerted on particle i due to its neighbor j from the pair potential Equation (1). The microscopic site energy (ϵ) is written as

$$\epsilon_i = \frac{1}{2}m_i|\mathbf{v}_i|^2 + \frac{1}{2}\sum_{j,j\neq i} u(r_{ij}), \tag{8}$$

where m is the mass of the particle. In MD simulations, the motion of particles is determined based on the integration of the equations of motion in subsequent simulation steps. To calculate the thermal conductivity, we discretized the right-hand side of Equation (6) in MD time steps as [48]

$$\lambda = \frac{\beta\Delta t}{VT}\sum_{l=1}^{M}\frac{1}{N_s - M}\sum_{k=1}^{N_s-l}\mathbf{j}(l+k)\mathbf{j}(k), \tag{9}$$

where N_s is the number of MD steps after equilibration and M is the number of steps over which the time average is calculated. M should be considerably smaller than the number of MD steps (N_s) to ensure good statistical averaging over time.

2.3. Simulation Details

MD simulations were performed in the NVT ensemble, where the temperature was controlled by velocity scaling [40]. That choice is based on previous studies in which velocity scaling and a Nosé–Hoover thermostat were used to control the temperature in order to determine the thermal conductivity using the Green–Kubo formula. Those studies showed that the temperature control method does not influence the obtained results that agree with each other [41]. The velocity Verlet algorithm was used to integrate the equations of motion. The integration time step was set to $\Delta t = 0.001\tau$ ($\tau = \sqrt{m\sigma^2/\varepsilon}$). In calculations, the following independent, dimensionless variables are used: the reduced number density $\rho^* = N\sigma^3/V$, the reduced pressure $p^* = p\beta\sigma^3$, and the reduced thermal conductivity $\lambda^* = \lambda\sigma\tau/k_B$. Simulations were performed for Yukawa systems based on the equation of state from Refs. [18,23] at the following parameters of contact potential $\beta\varepsilon$ = 20, 39, 81 and the inverse screening lengths $\kappa\sigma = 6.7, 10, 16.7$ at pressure $p^* = 60$. The present studies have been limited to the parameters of potential for which the fcc Yukawa crystals were observed and close to experimental conditions. For example, the electrostatic potential on the surface of an isolated particle embedded in a low dielectric solvent at room temperature is around 25 mV, which corresponds roughly to $\beta\varepsilon \approx 20$ [23]. To avoid phase transitions during simulation, the upper limit of the polydispersity parameter was taken $\delta = 0.08$ because the terminal polydispersity of studied systems for considered parameters of the Yukawa potential is around 0.095 (9.5%) [23]. The durations of the MD simulation runs were equal to 5×10^6 MD steps, after an equilibration of 10^5 MD steps. The number of steps M (Equation (9)) over which the time average in the Green–Kubo method was calculated was set to 10^5. All further presented thermal conductivity coefficients are values that were obtained for the thermodynamical limit ($N \to \infty$) using a linear approximation based on the results for four sizes of systems with a number of particles equal to N = 256, 500, 864, 4000. In addition, each thermal conductivity coefficient at a given thermodynamic condition was determined as the average of ten independent runs.

3. Results and Discussion

Charge-stabilized colloidal model crystals with various Debye screening lengths and contact potentials were simulated. We also studied the monodisperse systems in which the colloid particles have the same sizes, and the polydisperse systems in which the colloid particles have different sizes.

3.1. Thermal Conductivity of Monodisperse Yukawa Crystals

First, we calculated the thermal conductivity of the monodisperse Yukawa crystals for systems of different sizes and studied the finite-size effect (or size dependence) on the obtained results. We found a weak size dependence similar to that of other systems that were previously studied using the Green–Kubo approach in EMDs simulations [35,41,48,49]. It was observed that for a system of 256 particles, the thermal conductivity of the monodisperse Yukawa crystal differed from λ in the thermodynamic limit by only a few percent (see Figure 2). Regardless of the slight differences in values of thermal conductivities for finite systems and those in the thermodynamic limit, all the thermal conductivity coefficients presented below are given for values in the thermodynamic limit ($N \to \infty$).

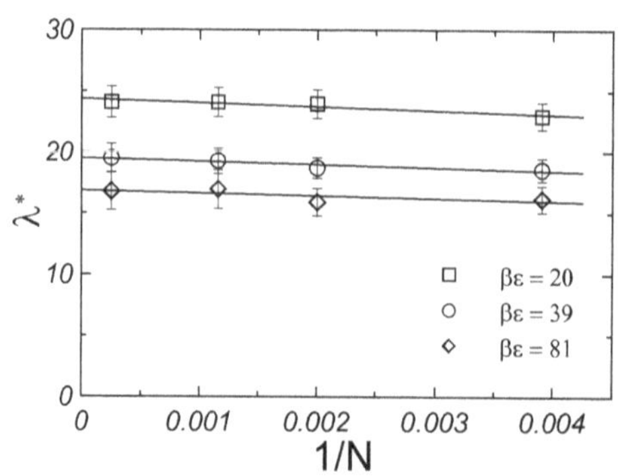

Figure 2. Thermal conductivity of monodisperse Yukawa crystals as a function $1/N$ at $\kappa\sigma = 10$ for various values of contact potential.

Figure 3 shows the dependence of the thermal conductivity of the monodisperse system on the parameters of the Yukawa potential. From Figure 3a, it can be seen that increasing the screening length, $(\kappa\sigma)^{-1}$, causes a decrease in the thermal conductivity of the system in a wide range of $\beta\varepsilon$. A similar dependence is observed in Figure 3b, where an increase in the value of parameter $\beta\epsilon$ causes a decrease in the thermal conductivity of the system. The observed dependences are partly related to the change in the density of the systems (for parameters of Yukawa potential $\beta\varepsilon$ and $\kappa\sigma$) presented in Figure 4. In the case of model colloidal crystals, increasing the system's density increases the thermal conductivity. Here, one can add that the parametrization of the Yukawa potential allows us to obtain similar thermal conductivity values for different contact potential values and screening lengths; for example, for $\kappa\sigma = 6.7$, $\beta\epsilon = 20$ and $\kappa\sigma = 16.7$, $\beta\epsilon = 81$ (see Figure 3). Following the equation of state [18,19] the monodisperse Yukawa systems for those two sets of parameters have similar densities, as shown in Figure 4.

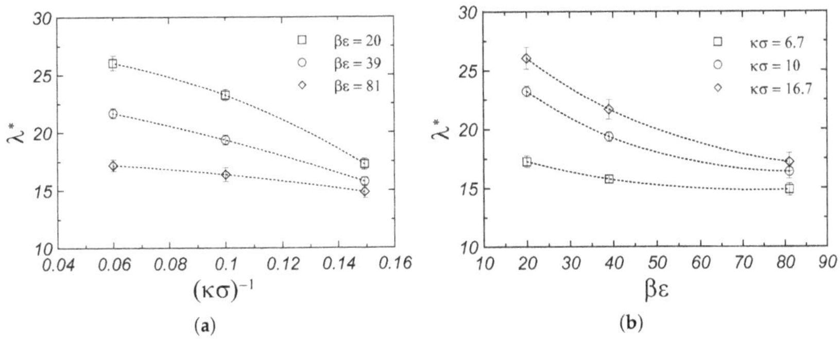

Figure 3. Thermal conductivity of monodisperse fcc Yukawa crystals: (**a**) as a function of Debye screening length and (**b**) as a function of contact potential. The results of the thermodynamic limit ($N \to \infty$) are presented. Dotted lines are drawn to guide the eyes.

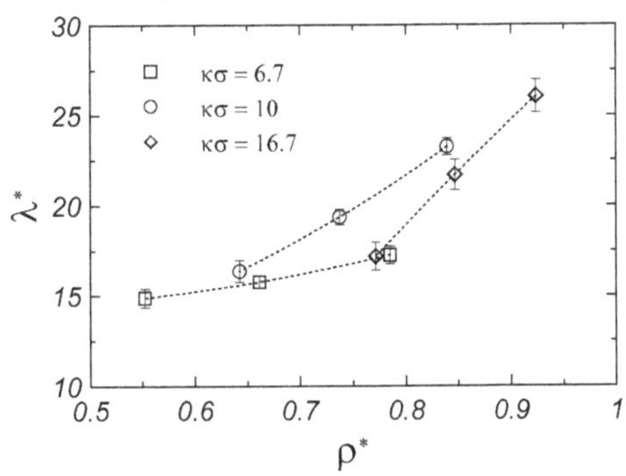

Figure 4. Density dependence of the thermal conductivity of monodisperse fcc Yukawa crystals. Dotted lines are drawn to guide the eyes.

Knowledge of the dependences of the thermal conductivity of the system on the parameters of the Yukawa potential is a reference point for studies regarding the effect of the particle size polydispersity on the thermal conductivity of polydisperse Yukawa crystals.

3.2. *Thermal Conductivity of Polydisperse Yukawa Crystals*

In real colloids, the particle sizes are slightly different. In our studies, this diversity is described by the polydispersity parameter δ (Equation (2)). Based on the phase diagrams determined by Dijkstra and co-authors [23,24] we considered Yukawa crystals with a polydispersity parameter value ranging from 0 to 8%, for which a face-centered cubic structure occurs.

Figure 5 shows the effect of particle size polydispersity, δ, on the thermal conductivity coefficient for all studied systems. In all cases, the effect of particle size polydispersity in the system on thermal conductivity is significant and causes a decrease in λ with increasing particle size polydispersity. The strongest effect of particle size polydispersity on the thermal conductivity of crystals is observed for systems with the shortest screening length (Figure 5a). For all systems, increasing the particle size polydispersity to $\delta = 8\%$ causes a four- to fivefold decrease in the thermal conductivity of fcc Yukawa crystals. It is worth noting that the largest considered particle size polydispersity (Figure 5) has a greater effect on the thermal conductivity of the system than the values of the contact potential and screening length in the considered monodisperse systems (Figure 3). Moreover, the change in density with an increase in the polydispersity of particle sizes is only 1% to 3%, depending on the screening length and contact potential. Therefore, the effect of the change in density in the polydisperse system probably makes a negligible contribution to such a strong reduction in thermal conductivity. Recent studies reveal the origin of the low-lattice thermal conductivity in all-inorganic halide perovskites $CsSnBr_{3\text{-}x}I_x$ [34]. Among other things, one reason for the low thermal conductivity in these materials is the dynamic structural fluctuations that come from a highly dynamic and disordered structure in $CsSnBr_{3\text{-}x}I_x$. Local disorder due to polydispersity of particle sizes may cause a similar effect, and this could be the subject of further investigation.

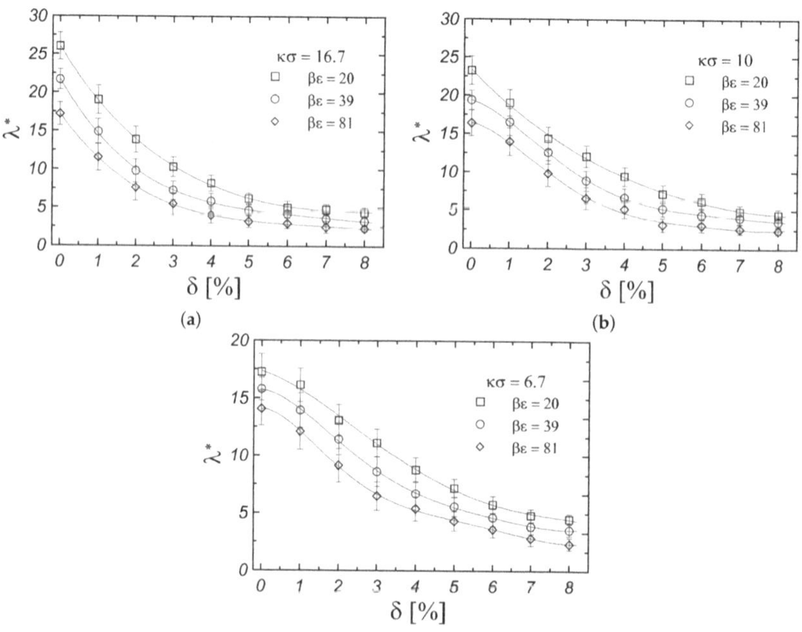

Figure 5. Thermal conductivity of fcc Yukawa crystals as a function of polydispersity for different values of contact potential and screening lengths $(\kappa\sigma)^{-1}$: (**a**) $\kappa\sigma = 16.7$, (**b**) $\kappa\sigma = 10$, (**c**) $\kappa\sigma = 6.7$. Solid lines are drawn to guide the eyes.

4. Conclusions

The thermal conductivities of monodisperse and polydisperse Yukawa crystals for a wide range of contact potential values and Debye screening lengths were determined by molecular dynamics simulations using the Green–Kubo method.

Studies involving monodisperse Yukawa crystals showed that an increase in the Debye screening length leads to a reduction in the thermal conductivity. For all the screening lengths considered, the thermal conductivity increases with the density of the crystal. The influence of the crystal density on λ is greater for the shorter screening lengths in the system. Extensive computer simulations have shown that the increase in the size polydispersity of particles in fcc Yukawa crystals leads to a strong decrease in their thermal conductivity. The present results qualitatively confirm recent experimental and theoretical reports on the effect of disorder on the reduction of the thermal conductivity coefficient of colloidal crystals (using a binary mixture as an example) [39]. However, it should be stressed that in the case of the binary mixture, a twofold decrease in the thermal conductivity coefficient associated with the global disorder observed in the glassy state has been found. On the other hand, in the case of polydisperse Yukawa crystals, about a fivefold decrease in the thermal conductivity coefficient was caused by a local disorder, though the regular crystal structure (fcc) was maintained. The origin of such large quantitative differences in the results may suggest the presence of different heat transport mechanisms (or a certain combination of common mechanisms) in polydisperse fcc Yukawa crystals and binary colloidal glass which may be related to local and global disorder, respectively. The above results motivate further research to determine the thermal conductivity coefficients of model amorphous colloidal systems and systems with other crystallographic structures, and special attention should be devoted to studying the mechanisms of heat transport in colloidal crystals. It seems particularly interesting to take into account the variation in particle sizes in these models.

In closing, it is worth mentioning that phononic crystals can control the transport of heat and sound similar to the control of electric currents by semiconductors or light by photonic crystals. It is known that various phononic crystals can be fabricated using self-assembling submicrometer particles into colloidal crystals [50]. Here, it should be stressed that the main advantages of colloidal crystals are their large volume, low cost, and their ability to maintain high-quality translational order [50]. That indicates the importance of the presented studies for a deeper understanding of heat transport processes in colloidal crystals from a theoretical and practical point of view.

Author Contributions: Conceptualization, K.V.T. and K.H.; methodology, K.V.T.; software, K.H.; validation, K.H. and K.V.T.; formal analysis, K.H. and K.V.T.; investigation, K.H.; resources, K.V.T.; data curation, K.H.; writing—original draft preparation, K.V.T.; writing—review and editing, K.V.T.; visualization, K.H.; supervision, K.V.T.; project administration, K.V.T.; funding acquisition, K.V.T. All authors have read and agreed to the published version of the manuscript.

Funding: This work was partially supported by the grant No. 2017/27/B/ST3/02955 of the National Science Centre, Poland.

Institutional Review Board Statement: Not applicable.

Informed Consent Statement: Not applicable.

Data Availability Statement: Data is contained within the article.

Acknowledgments: The authors thank Krzysztof W. Wojciechowski for reading this manuscript and providing useful comments. The computations were partially performed at Poznań Supercomputing and Networking Center (PCSS).

Conflicts of Interest: The authors declare no conflicts of interest. The funders had no role in the design of the study; in the collection, analyses, or interpretation of data; in the writing of the manuscript; or in the decision to publish the results.

Abbreviations

The following abbreviations are used in this manuscript:

MDs	Molecular dynamics
EMD	Equilibrium molecular dynamics
HCRYP	Hard-core repulsive Yukawa potential
fcc	Face-centered cubic

References

1. van Blaaderen, A.; Ruel, R.; Wiltzius, P. Template-directed colloidal crystallization. *Nature* **1997**, *385*, 321–324.

2. Talapin, D.V.; Lee, J.S.; Kovalenko, M.V.; Shevchenko, E.V. Prospects of colloidal nanocrystals for electronic and optoelectronic applications. *Chem. Rev.* **2010**, *110*, 389–458. [PubMed]
3. Galisteo-López, J.; Ibisate, M.; Sapienza, R.; Froufe-Peérez, L.; Blanco, A.; López, C. Self-Assembled Photonic Structures. *Adv. Mater.* **2011**, *23*, 30. [PubMed]
4. Alexeev, V.L.; Finegold, D.N.; Asher, S.A. Photonic crystal glucose-sensing material for noninvasive monitoring of glucose in tear fluid. *Clin. Chem.* **2004**, *50*, 2353–2360. [PubMed]
5. Demirörs, A.F.; Johnson, P.M.; van Kats, C.M.; van Blaaderen, A.; Imhof, A. Directed Self-Assembly of Colloidal Dumbbells with an Electric Field. *Langmuir* **2010**, *26*, 14466–14471. [CrossRef] [PubMed]
6. Bakker, H.E.; Dussi, S.; Droste, B.L.; Besseling, T.H.; Kennedy, C.L.; Wiegant, E.I.; Liu, B.; Imhof, A.; Dijkstra, M.; van Blaaderen, A. Phase diagram of binary colloidal rod-sphere mixtures from a 3D real-space analysis of sedimentation—Diffusion equilibria. *Soft Matter* **2016**, *12*, 9238–9245. [CrossRef]
7. Smallenburg, F.; Vutukuri, H.R.; Imhof, A.; van Blaaderen, A.; Dijkstra, M. Self-assembly of colloidal particles into strings in a homogeneous external electric or magnetic field. *J. Phys.-Condes. Matter* **2012**, *24*, 464113.
8. Demirörs, A.F.; Pillai, P.P.; Kowalczyk, B.; Grzybowski, B.A. Colloidal assembly directed by virtual magnetic moulds. *Nature* **2013**, *503*, 99–103. [CrossRef]
9. Sacanna, S.; Rossi, L.; Pine, D.J. Magnetic Click Colloidal Assembly. *J. Am. Chem. Soc.* **2012**, *134*, 6112–6115.
10. Likos, C. Effective interactions in soft condensed matter physics. *Phys. Rep.-Rev. Sec. Phys. Lett.* **2001**, *348*, 267–439. [CrossRef]
11. Verwey, E.J.W.; Overbeek, J.T.G. *Theory of the Stability of Lyophobic Colloids*; Elsevier: New York, NY, USA, 1948.
12. Alexander, S.; Chaikin, P.; Grant, P.; Morales, G.; Pincus, P.; Hone, D. Charge renormalization, osmotic-pressure, and bulk modulus of colloidal crystals: Theory. *J. Chem. Phys.* **1984**, *80*, 5776–5781. [CrossRef]
13. Ohshima, H. *Theory of Colloid and Interfacial Electric Phenomena*; Academic Press: Amsterdam, The Netherlands, 2006.
14. Hone, D.; Alexander, S.; Chaikin, P.M.; Pincus, P. The phase-diagram of charged colloidal suspensions. *J. Chem. Phys.* **1983**, *79*, 1474. [CrossRef]
15. Tejero, C.F.; Lutsko, J.F.; Colot, J.L.; Baus, M. Thermodynamic properties of the fluid, FCC, and BCC phases of monodisperse charge-stabilized colloidal suspensions within the Yukawa model. *Phys. Rev. A* **1992**, *46*, 3373. [CrossRef] [PubMed]
16. Royall, C.; Leunissen, M.; Hynninen, A.; Dijkstra, M.; van Blaaderen, A. Re-entrant melting and freezing in a model system of charged colloids. *J. Chem. Phys.* **2006**, *124*, 244706. [CrossRef] [PubMed]
17. Azhar, F.E.; Baus, M.; Ryckaert, J.P. Line of triple points for the hard-core Yukawa model: A computer simulation study. *J. Chem. Phys.* **2000**, *112*, 5121–5126. [CrossRef]
18. Hynninen, A.P.; Dijkstra, M. Phase diagrams of hard-core repulsive Yukawa particles. *Phys. Rev. E* **2003**, *68*, 021407. [CrossRef]
19. Tretiakov, K.V.; Wojciechowski, K.W. Partially auxetic behavior in fcc crystals of hard-core repulsive Yukawa particles. *Phys. Stat. Solidi B* **2014**, *251*, 383. [CrossRef]
20. Tretiakov, K.V.; Wojciechowski, K.W. The Influence of the Soft Yukawa Potential and Hard Core Interactions on Auxeticity of the Face Centered Cubic Crystal of Hard-Core Repulsive Yukawa Particles. *Phys. Stat. Solidi B* **2020**, *257*, 2000194. [CrossRef]
21. Phan, S.E.; Russel, W.B.; Zhu, J.X.; Chaikin, P.M. Effects of polydispersity on hard sphere crystals. *J. Chem. Phys.* **1998**, *108*, 9789.
22. Phan, S.E.; Russel, W.B.; Cheng, Z.; Zhu, J.; Chaikin, P.M.; Dunsmuir, J.H.; Ottewill, R.H. Phase transition, equation of state, and limiting shear viscosities of hard sphere dispersions. *Phys. Rev. Stat. Phys. Plasmas Fluids Relat. Interdiscip. Top.* **1996**, *54*, 6633–6645. [CrossRef]
23. Colombo, J.; Dijkstra, M. Effect of quenched size polydispersity on the fluid-solid transition in charged colloidal suspensions. *J. Chem. Phys.* **2011**, *134*, 154504. [CrossRef] [PubMed]
24. van der Linden, M.N.; van Blaaderen, A.; Dijkstra, M. Effect of size polydispersity on the crystal-fluid and crystal-glass transition in hard-core repulsive Yukawa systems. *J. Chem. Phys.* **2013**, *138*. [CrossRef] [PubMed]
25. Narojczyk, J.W.; Piglowski, P.M.; Wojciechowski, K.W.; Tretiakov, K.V. Elastic properties of mono- and polydisperse two-dimensional crystals of hard-core repulsive Yukawa particles. *Phys. Stat. Solidi B* **2015**, *252*, 1508–1513. [CrossRef]
26. Tretiakov, K.V.; Wojciechowski, K.W. Elasticity of two-dimensional crystals of polydisperse hard disks near close packing: Surprising behavior of the Poisson's ratio *J. Chem. Phys.* **2012**, *136*, 204506. [CrossRef] [PubMed]
27. Tretiakov, K.V.; Wojciechowski, K.W. Elastic properties of fcc crystals of polydisperse soft spheres. *Phys. Stat. Solidi B* **2013**, *250*. [CrossRef]
28. Piglowski, P.; Narojczyk, J.; Wojciechowski, K.; Tretiakov, K. Auxeticity enhancement due to size polydispersity in fcc crystals of hard-core repulsive Yukawa particles. *Soft Matter* **2017**, *13*, 7916. [CrossRef]
29. Cahill, D.; Ford, W.; Goodson, K.; Mahan, G.; Majumdar, A.; Maris, H.; Merlin, R.; Phillpot, S. Nanoscale thermal transport. *J. Appl. Phys.* **2003**, *93*, 793–818. [CrossRef]
30. Cahill, D.G.; Braun, P.V.; Chen, G.; Clarke, D.R.; Fan, S.; Goodson, K.E.; Keblinski, P.; King, W.P.; Mahan, G.D.; Majumdar, A.; et al. Nanoscale thermal transport. II. 2003–2012. *Appl. Phys. Rev.* **2014**, *1*, 011305. [CrossRef]
31. Poudel, B.; Hao, Q.; Ma, Y.; Lan, Y.; Minnich, A.; Yu, B.; Yan, X.; Wang, D.; Muto, A.; Vashaee, D.; et al. High-thermoelectric performance of nanostructured bismuth antimony telluride bulk alloys. *Science* **2008**, *320*, 634–638. [CrossRef]
32. Cahill, D.; Watson, S.; Pohl, R. Lower limit to the thermal-conductivity of disordered crystals. *Phys. Rev. B* **1992**, *46*, 6131–6140. [CrossRef]

33. Ishibe, T.; Kaneko, T.; Uematsu, Y.; Sato-Akaba, H.; Komura, M.; Iyoda, T.; Nakamura, Y. Tunable Thermal Switch via Order-Order Transition in Liquid Crystalline Block Copolymer. *Nano Lett.* **2022**, *22*, 6105–6111. [CrossRef] [PubMed]
34. Xie, H.; Hao, S.; Bao, J.; Slade, T.J.; Snyder, G.J.; Wolverton, C.; Kanatzidis, M.G. All-Inorganic Halide Perovskites as Potential Thermoelectric Materials: Dynamic Cation off-Centering Induces Ultralow Thermal Conductivity. *J. Am. Chem. Soc.* **2020**, *142*, 9553–9563. [CrossRef] [PubMed]
35. Tretiakov, K.V.; Hyzorek, K. Role of the phonon confinement effect and boundary scattering in reducing the thermal conductivity of argon nanowire. *J. Chem. Phys.* **2021**, *154*, 054702.
36. Nutz, F.A.; Ruckdeschel, P.; Retsch, M. Polystyrene colloidal crystals: Interface controlled thermal conductivity in an open-porous mesoparticle superstructure. *J. Colloid Interface Sci.* **2015**, *457*, 96–101. [CrossRef] [PubMed]
37. Nutz, F.A.; Retsch, M. Interfacial and volumetric sensitivity of the dry sintering process of polymer colloidal crystals: A thermal transport and photonic bandgap study. *Phys. Chem. Chem. Phys.* **2017**, *19*, 16124–16130. [CrossRef]
38. Still, T.; Cheng, W.; Retsch, M.; Sainidou, R.; Wang, J.; Jonas, U.; Stefanou, N.; Fytas, G. Simultaneous occurrence of structure-directed and particle-resonance-induced phononic gaps in colloidal films. *Phys. Rev. Lett.* **2008**, *100*, 194301. [CrossRef]
39. Nutz, F.A.; Philipp, A.; Kopera, B.A.F.; Dulle, M.; Retsch, M. Low Thermal Conductivity through Dense Particle Packings with Optimum Disorder. *Adv. Mater.* **2018**, *30*, 1704910. [CrossRef]
40. Allen, M.P.; Tildesley, D.J. *Computer Simulation of Liquids*; J. W. Arrowsmith Ltd.: Bristol, UK, 1987.
41. Hyżorek, K.; Tretiakov, K. Thermal conductivity of liquid argon in nanochannels from molecular dynamics simulations. *J. Chem. Phys.* **2016**, *144*, 194507. [CrossRef]
42. Russel, W.B.; Saville, D.A.; Schowalter, W.R. *Colloidal Dispersions*; Cambridge University Press: Cambridge, UK, 1992.
43. Green, M.S. Markoff Random Processes and the Statistical Mechanics of Time-Dependent Phenomena. II. Irreversible Processes in Fluids. *J. Chem. Phys.* **1954**, *22*, 398–413. [CrossRef]
44. Kubo, R. Statistical-Mechanical Theory of Irreversible Processes. I. General Theory and Simple Applications to Magnetic and Conduction Problems. *J. Phys. Soc. Jpn.* **1957**, *12*, 570–586.
45. Zwanzig, R. Time-Correlation Functions and Transport Coefficients in Statistical Mechanics. *Annu. Rev. Phys. Chem.* **1965**, *16*, 67–102. [CrossRef]
46. Hansen, J.P.; McDonald, I.R. *Theory of Simple Liquids*; Academic: New York, NY, USA, 2005.
47. Heyes, D. Transport coefficients of Lennard-Jones fluids: A molecular-dynamics and effective-hard-sphere treatment. *Phys. Rev. B* **1988**, *37*, 5677. [CrossRef] [PubMed]
48. Tretiakov, K.V.; Scandolo, S. Thermal conductivity of solid argon for molecular dynamics simulations. *J. Chem. Phys.* **2004**, *120*, 3765–3769. [CrossRef] [PubMed]
49. Vogelsang, R.; Hoheisel, C.; Ciccotti, G. Thermal conductivity of the Lennard-Jones liquid by molecular dynamics calculations. *J. Chem. Phys.* **1987**, *86*, 6371. [CrossRef]
50. Vasileiadis, T.; Varghese, J.; Babacic, V.; Gomis-Bresco, J.; Navarro Urrios, D.; Graczykowski, B. Progress and perspectives on phononic crystals. *J. Appl. Phys.* **2021**, *129*, 160901. [CrossRef]

materials

MDPI

Article

Theoretical Insights into Twist–Bend Nematic Liquid Crystals: Infrared Spectra Analysis of Naphthalene-Based Dimers

Barbara Loska [1], Yuki Arakawa [2] and Katarzyna Merkel [1,*]

[1] Institute of Materials Engineering, Faculty of Science and Technology, University of Silesia in Katowice, ul. 75 Pułku Piechoty 1A, 41-500 Chorzów, Poland; barbara.loska@us.edu.pl
[2] Department of Applied Chemistry and Life Science, Graduate School of Engineering, Toyohashi University of Technology, 1-1 Hibarigaoka, Tempaku-cho, Toyohashi 441-8580, Aichi, Japan; arakawa@tut.jp
* Correspondence: katarzyna.merkel@us.edu.pl

Abstract: In this study, we employed density functional theory (DFT), a standard method in quantum chemistry, to investigate the structural intricacies of thioether-linked naphthalene-based liquid-crystal dimers. The theoretical analysis included the calculation of the molecular bend angle, a crucial factor influencing the formation of the twist–bend nematic (N_{TB}) phase, as well as other molecular parameters such as transition dipole moments, bond lengths, and bond energies. These calculations allowed for the determination of the probable conformations and the computation of their vibrational spectra, which are essential for interpreting experimental spectra. Connecting these insights, we identified stable conformations and observed differences in the spectra between the conventional nematic (N) and N_{TB} phases. The combined DFT calculations and infrared absorbance measurements allowed us to investigate the structure and intermolecular interactions of molecules in the N and N_{TB} phases of the dimers. Notably, significant changes in average absorbance were detected in the experimental spectra in the N_{TB} phase. During the transition from the N phase to the N_{TB} phase, a clear decrease in absorbance for longitudinal dipoles and an increase for transverse dipoles were observed. This phenomenon suggests that longitudinal dipoles are antiparallel, while transverse dipoles are parallel. To verify the influence of nearest-neighbor interactions, DFT calculations were conducted on a system comprising several neighboring molecules.

Keywords: FTIR spectroscopy; DFT simulations; conformational probability; liquid crystal dimers; spatially modulated phases; twist–bend nematics

1. Introduction

Liquid crystals occupy a unique realm between isotropic liquids and crystalline solids, exhibiting remarkable properties that have garnered significant interest across various scientific disciplines. Among them, a twist–bend nematic (N_{TB}) phase has recently emerged as an intriguing class. The discovery of the N_{TB} phase involved a series of significant contributions. The concept of spontaneous symmetry breaking in mesophases of achiral bent-shaped molecules was initially explored by Meyer in 1973 [1]. This theoretical groundwork laid the foundation for further investigations, which would come a couple of decades later. In 2001, Dozov examined spontaneous symmetry breaking in mesophases of achiral bent-shaped molecules and provided insights into the potential existence of novel liquid-crystal phases, characterized by a negative bend elastic constant [2]. In subsequent investigations, Memmer conducted computer simulation studies in 2002 to explore the lower-symmetry liquid-crystal phases Dozov anticipated, further supporting the existence of these distinct mesophases and providing valuable insights into their characteristics [3].

It took nearly another decade for the experimental confirmation of this intriguing phase to take place. In 2010, Panov et al. provided compelling experimental evidence for the existence of the N_{TB} phase in nonchiral planar-aligned bimesogens [4]. Their groundbreaking work revealed spontaneous periodic deformations and a transition from a nematic (N) to another N phase accompanied by a negative elastic constant, thus confirming the presence of the N_{TB} phase and its distinct characteristics. This experimental verification opened up new avenues of exploration and prompted further investigations into the twist–bend phenomenon. Subsequently, in 2011, Cestari et al. conducted a comprehensive study focusing on the liquid-crystal dimer 1″,7″-bis(4-cyanobiphenyl-4′-yl) heptane (CB7CB) [5]. Their investigation shed further light on the N_{TB} phase, providing detailed insights into its phase behavior and properties.

Another decade later, we are continuing the exploration of twist–bend nematogens, actively searching for new liquid-crystal dimers that exhibit this intriguing phase. N_{TB} phases exhibit a modulated N structure, characterized by the presence of helical arrangements on a nanoscale, with a pitch length spanning a few nanometers [6–8]. These helical structures give rise to fascinating optical properties and have drawn considerable attention for their potential applications in advanced display technologies and photonic devices [9–11].

While the N_{TB} assignment has been widely adopted to describe the lower-temperature-modulated nematic phase found in bent-shaped dimers such as CB7CB, it is worth noting that this interpretation remains a subject of ongoing debate. In particular, Vanakaras and co-authors have highlighted that the observed nanoscale pitch modulation may not correspond to the twist–bend structure originally envisioned by Meyer [12]. Instead, it has been argued that the observed features of the so-called N_X phase are more consistent with a polar-twisted nematic (N_{PT}) structure, as described in the theoretical model proposed by Vanakaras and Photinos [13–16]. While many studies—including the present one—follow the widely accepted N_{TB} nomenclature for clarity and consistency with the earlier literature, the nature of this modulated nematic phase is still under discussion.

In this research, through computational techniques employing density functional theory (DFT), we aim to gain insights into the electronic structure, molecular configurations, and vibrational properties of these intriguing materials. Understanding the fundamental aspects of twist–bend nematics is vital for tailoring their properties and designing novel liquid-crystal materials with enhanced performance. It is widely acknowledged that the formation of the N_{TB} phase necessitates a molecular bending. Therefore, the assessment of conformational probabilities in N_{TB}-forming molecules has emerged as an indispensable approach [17–20]. This is because a singular conformation at the energy minimum is insufficient for comprehensively evaluating molecular curvature and other attributes in flexible liquid-crystal dimers.

Through this paper, we contribute to the broader understanding of twist–bend nematics and advance the knowledge of liquid-crystal materials, opening new avenues for their application in various technological fields. In this study, we present a theoretical and spectroscopic investigation of thioether-linked 6-(4-cyanophenyl)-2-naphtalene-based dimers with the heptyl and the propyl chain in the spacer. The spacer contains an odd number of methylene units, similarly to cyanobiphenyl-based dimers like the archetypal N_{TB}-forming CB7CB and its homologs, which have been extensively studied [5,6,8,21–26]. Thioether-linked CBS*n*SCB dimers exhibited a smaller dimer bend angle compared to methylene- and ether- linked molecules, facilitating a shorter pitch length of the N_{TB}-phase helical formation. Consequently, the dimers examined in this study, denoted as (CN)PNS*n*SNP(CN), feature a thioether linkage; however, in the biphenyl mesogen, the inner phenyl was exchanged for naphthalene. A related molecular design incorporating a naphthalene core has also been reported in the form of benzoyloxynaphthyl-based dimers, which showed comparable liquid crystalline behavior [27]. By employing the polarized infrared (IR) absorbance method on a homogeneously aligned sample, we obtained valuable insights into the orientation of the molecules' transition dipole moments. These findings were then compared with the theoretically calculated Cartesian components of the vibrational transition dipole moment for specific vibrations, thereby facilitating precise band assignment in the experimental spectra, which allowed for the analysis of the IR spectroscopy measurements.

2. Materials and Methods

2.1. Materials

In this research, we focused on the investigation of thioether-linked dimers based on the 6-(4-cyanophenyl)-2-naphthyl- rigid core, with the specific dimer acronym (CN)PNS*n*SNP(CN), where PN denotes a phenyl-naphthalene ring core, (CN) is the cyano group, and n is the odd number of carbons in the alkyl spacer. The terminal groups of the dimer are chemically linked to each other through thioether bridges. Thioether-linked dimers, which incorporate asymmetric π-conjugated mesogenic arms with terminal cyano groups, have been shown to offer significant advantages in realizing materials that exhibit a diverse range of N_{TB} phases and glassy N_{TB} states even at room temperature. Synthesis details and a preliminary DSC investigation of the compounds have been published by Arakawa et al. [28]. Figure 1 illustrates the chemical structure and transition temperatures of the studied dimers determined by Arakawa et al. by means of Differential Scanning Calorimetry (DSC).

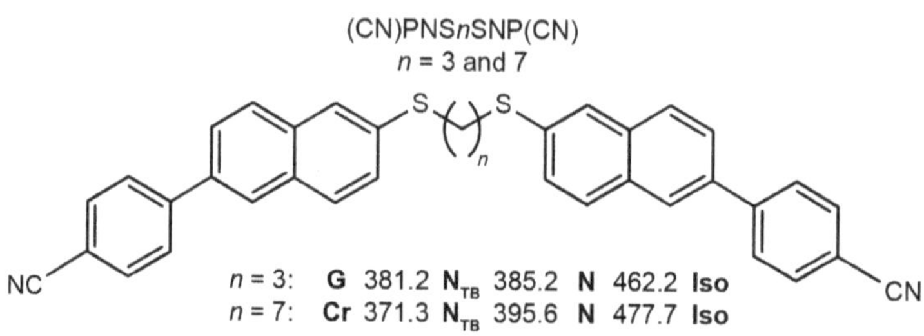

Figure 1. Chemical structure of the studied dimers and their phase-transition temperatures.

2.2. Polarizing Optical Microscopy

Polarizing Optical Microscopy (POM) was employed to evaluate the alignment quality of the prepared ZnSe cells (Crystan, Dorset, UK), as well as for the preliminary identification of mesophases and determination of phase-transition temperatures. The liquid-crystal textures of the samples were observed using a polarizing microscope (Olympus BX56, Olympus, Tokyo, Japan) equipped with a PID-controlled temperature system, ensuring an accuracy of 2 mK, and are presented in Figure 2.

2.3. IR Measurements

The IR measurement cells used in this study achieved planar alignment by sandwiching the dimer material between two optically polished zinc selenide (ZnSe) disks. These disks were spin-coated with an SE-130 commercial polymer aligning agent (Nissan Chemical Industries, Ltd., Pasadena, TX, USA). The aligning agent was rubbed after polymerization and the cell was assembled with an antiparallel rubbing direction to achieve a planar alignment in the sample. The thickness of the fabricated cells was achieved by using a Mylar foil spacer (GoodFellow Ltd., Huntingdon, UK) and determined, by measuring interference fringes using a spectrometer interfaced with a PC (Avaspec-2048, Avantes, Apeldoorn, The Netherlands), to be 5.62 μm and 5.77 μm for the shorter (n = 3) and longer (n = 7) studied dimers, respectively.

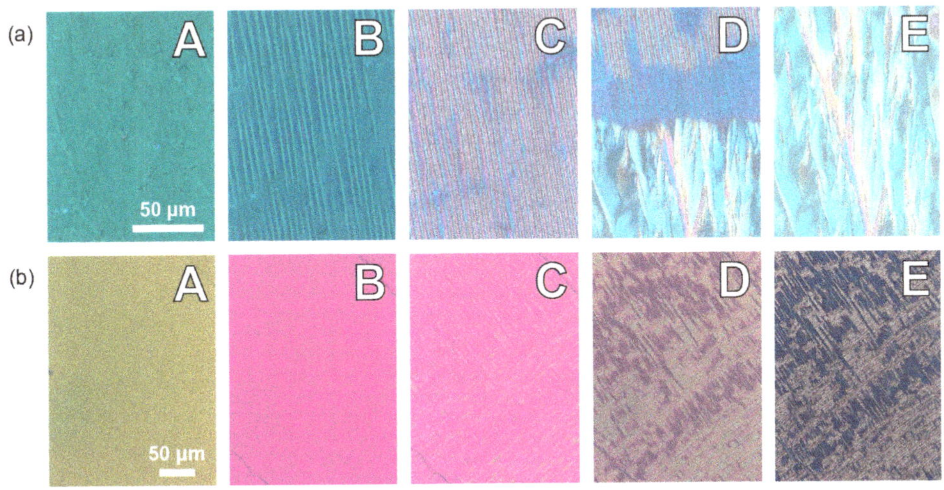

Figure 2. Textures observed during cooling at a rate of 2 K/min: (**a**) (CN)PNS7SNP(CN) sample: A—N phase (445 K); B—N–N_{TB} transition (396 K); C—N_{TB} phase (392 K); D—N_{TB}-Cr transition (390 K); E—Cr phase (385 K). (**b**) (CN)PNS3SNP(CN) sample: A—N phase (450 K); B—N–N_{TB} transition (378 K); C—N_{TB} phase (373 K); D and E—N_{TB} glass (363 K and 310 K, respectively). All textures were observed using ZnSe cells with a thickness of 5 μm.

IR spectra were obtained using an Agilent Cary 670 FTIR Fourier IR spectrometer (Agilent Technologies, Santa Clara, CA, USA). The transmission method with a polarized IR beam was employed for the experimental setup, where an IR-KRS5 grid polarizer (Specac Ltd., Orpington, UK) was utilized to polarize the IR beam. The acquired IR spectra were measured across the wavenumber range of 500–4000 cm^{-1}. These measurements facilitated the determination of the orientation of transition dipole moments with respect to the long molecular axis and the temperature-dependent behavior of sample absorbance.

Generally, to determine all three components of absorbance (A_x, A_y, A_z), two samples with different orientations are required: planar (homogeneous) and homeotropic. However, achieving a satisfactory homeotropic alignment proved challenging for the tested cyanophenyl-napthalene dimer materials, due to the chemical composition of the terminal groups of the dimer. Consequently, assuming the uniaxial nature of the material, A_x was assumed to be equal to A_y, and the mean absorbance (A_0) was calculated as $A_0 = (2A_y + A_z)/3$. The absorbance components were quantified by measuring the area bound by the contour of a specific band using Bio-Rad Win-IR Pro version 2.96e.

2.4. DFT Calculations

Electronic structure calculations of the molecules were performed using the Gaussian16 software package [29]. Density functional theory (DFT) was employed to calculate various molecular properties, including molecular structures, harmonic vibrational force constants, absolute IR intensities, and components of transition dipole moments. The B3LYP functional, which combines Becke's three-parameter exchange functional with the

Lee, Yang, and Parr correlation functional, was utilized. The polarization basis set (6-311G (d,p)) was used for the calculations [30,31].

To calculate the spectral density components, information about the transition dipole moments for specific vibrations was used. The parallel component of the absorption coefficient was determined by squaring the component of the transition dipole moment along the axis coinciding with the long axis of the dimer ($|\mu_z|^2 = \mu^2_{\|}$). Similarly, the perpendicular component of the spectral density was determined by summing the squares of the transition dipole moments along the vertical directions ($|\mu_x|^2 + |\mu_y|^2 = \mu^2_{\perp}$). The direction of the transition dipole moment was determined according to the molecular reference system.

Theoretical vibrational frequencies were appropriately scaled. According to the Computational Chemistry Comparison and Benchmark Database, for the polarization base and method used, the scaling coefficient was determined to be 0.967 ± 0.021. We found that two scaling factors facilitated better comparability with experimental results, and thus we split the data into two regions: below and above 2000 cm^{-1}. We empirically determined the coefficients for these ranges to be 0.9519 and 0.9736, respectively. Both of these values fall within the error margin of the database values. Gaussian profiles with a full width at half maximum (FWHM) of 7 cm^{-1} were employed. The visualization of the results was performed using GaussView 6 [32].

We first calculated the rotational potential barriers of the (CN)PNS3SNP(CN) molecules, using the relaxed potential energy surface scan method with molecular geometry optimization. Relaxed scans about the φ_1 of the molecule angles ($36 \times 10^\circ$ steps) while varying the fixed φ_2 angles ($18 \times 10^\circ$ steps) were performed. As the energy barrier to the internal rotation in the alkyl chain is very small (approx. 1 kJ/mol), only the all-trans conformation of the linker was considered. Rotational potential barriers were calculated considering the minimum energy of the whole population of dimers to create a 3D energy barrier map. Subsequently, based on the Cartesian coordinates of the lowest-energy geometries, the angle between the two mesogenic units was calculated.

3. Results

3.1. Energy Barrier Map and Conformers

The 3D plot of the rotational potential (energy barrier) map for all considered conformers is presented in Figure 3. The dihedral energy of the absolute minimum among all scans was taken equal to zero. Because of local symmetry, we can identify three distinct minima:

- The symmetrical conformation of minimum energy $\varphi_1 = \varphi_2 = 0^\circ$ (referred to as the flat conformation) with energy barrier $\Delta E = 7.79$ kJ/mol;
- The asymmetrical conformation $\varphi_1 = 0$ and $\varphi_2 = 110^\circ$, with energy barrier $\Delta E = 7.31$ kJ/mol (referred to as the mixed conformation);
- The symmetrical conformation where $\varphi_1 = \varphi_2 = 110^\circ$—with an energy barrier of $\Delta E = 6.19$ kJ/mol (referred to as the upright conformation).

For further calculations, only the dimers corresponding to the lowest-energy conformations were investigated. Given the small number of alkyl groups in the linker, only the *all-trans* conformation was considered. Based on the Cartesian coordinates of the atoms of the calculated conformers, we found the bend angles of the molecules and calculated the Boltzmann distribution to obtain the probability-weighted average bend angles of the molecules.

Through the use of the energies of the calculated conformers, the probability-weighted average opening angle θ_{avg} was calculated to be 100.8°. The introduction of the weighted average based on the Boltzmann distribution to the bend angle calculation did not yield substantial variations in the resulting bend angle values; the observed variance remained within the range of 1.3°. Similar calculations were performed for dimers such as CB9CB, CBS7SCB, and CBS7OCB. The obtained results include the minimum and average values of the bend angles: CB9CB ($\theta_{min} = 111.4$; $\theta_{avg} = 103.7$), CBS7SCB ($\theta_{min} = 94.1$; $\theta_{avg} = 79.1$), and CBS7OCB ($\theta_{min} = 127.4$; $\theta_{avg} = 114.1$). In these calculations, the conformations of alkyl chains in the spacer were also taken into account [17–20].

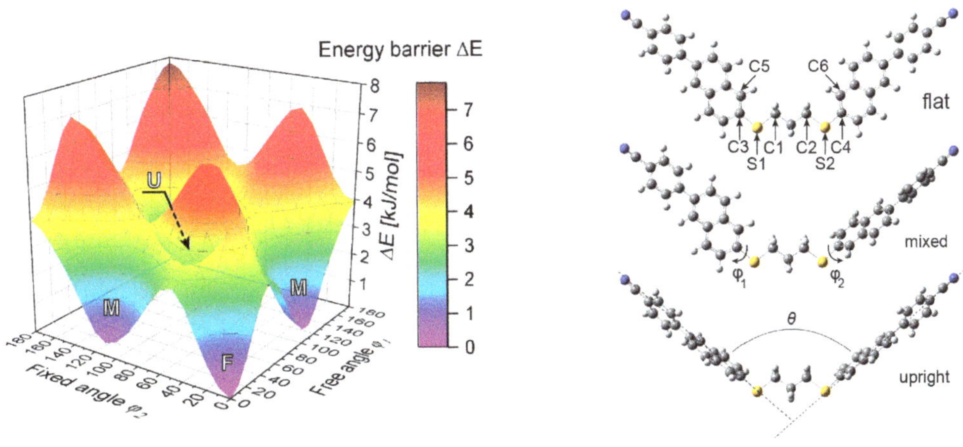

(a) (b)

Figure 3. (a) A 3D plot of the energy barriers for calculated dimer conformations. M—mixed; F—flat; U—upright conformers. (b) The identified minimum-energy conformations: top—the flat conformation with atom number labels; middle—the mixed conformation with torsional angle designations; bottom—the upright conformation and opening angle. Table 1 shows a comparison of the minimum-energy bend angles θ_{min} and the probability-weighted average bend angle θ_{avg} for the simulated (CN)PNS3SNP(CN) molecule.

It is important to note that our calculations did not account for the presence of hairpin conformers. In a real sample, these conformers might lower the average molecular bend. However, determining bend angles for isolated molecules is insufficient for a direct study of the conformational population.

Table 1. Energy barrier, ΔE, and molecular bend angles for the global energy minimum conformation, θ_{min}, of the calculated (CN)PNS3SNP(CN) molecule.

Conformation	Torsional Angles	Potential Energy $\left[\frac{\text{kJ}}{\text{mol}}\right]$	θ_{min} [°]
Flat	$\varphi_1 = \varphi_2 = 0°$	−6,122,426.6725	99.3
Mixed	$\varphi_1 = 0°$ $\varphi_2 = 110°$	−6,122,426.1981	99.5
Upright	$\varphi_1 = \varphi_2 = 110°$	−6,122,425.1571	100.3

3.2. Vibrational Spectra Band Assignments

Achieving perfect experimental order is inherently challenging, and a multitude of factors, particularly intermolecular interactions, influence the dichroism of spectral bands. Consequently, solely relying on the analysis of experimental spectra makes it difficult to definitively pinpoint which band and which para-axis of the dimer components correspond to the behavior of the dimer's long axis.

For symmetric dimers like the ones studied, each band corresponding to the vibration of the phenyl-naphthalene rigid core is anticipated to yield two characteristic vibrations with similar frequencies but differing intensities. These vibrations, occurring in both arms and characterized by one being in phase while the other is out of phase, can interact and result in a transition dipole moment either along or across the dimer's long axis. An analysis of band dichroism in the theoretical spectra, representing an ideal order, suggests that if "infinite dichroism" is attained—where one component reaches maximum intensity while the other nears zero—such a band describes the behavior of the dimer's long axis. Conversely, when band dichroism falls in the intermediate range, indicating a lack of vibration coupling in both arms, this band characterizes the behavior of the mesogens' para-axes.

Generally, the spectra can be divided into frequency ranges, each associated with specific vibrational characteristics:

- 500–950 cm^{-1}—these encompass deformational vibrations involving the carbon atoms (C-C) and involving linker sulfur atoms (C-S-C), as well as deformations of the hydrogen atoms (C-H) oriented out of the plane of the rigid cores of the dimer;
- 1000–1650 cm^{-1}—this range encompasses a broad spectrum of vibrational modes, including characteristic deformations within the naphthalene and benzene plane and deformations of the methylene groups within the alkyl chain linker of the dimer;
- 2100–2400 cm^{-1}—this range notably features the stretching vibrations of the cyan group (C-N), manifested as a sharp and highly intense peak in the experimental spectrum;
- 2900–3100 cm^{-1}—encompassing stretching vibrations of hydrogen atoms (C-H) within the aromatic ring, this range exhibits challenges in theoretical reproduction. The spectral bands in this range pertain to mixed vibrations characterized by significant overlap and distortions, primarily influenced by the Fermi resonance effect.

Table 2 provides a summary of the primary experimental bands observed, along with their assignments to specific vibrational modes of the functional groups. A detailed analysis of the polarized spectra, transition dipole moments, and their comparison with the theoretical spectrum for the most energetically stable conformation was conducted for the molecule with a seven-membered chain and is provided in Table S1 in the Supplementary Materials.

Table 2. Vibrational frequencies, IR intensities, and assignments of the most important bands in the (CN)PNS*n*SNP(CN) dimer series.

$\nu\,[cm^{-1}]$	I_{rel}	Assignment	μ
570	w	γCC op rc + $\delta C_{Ar}S$ + δCN	⊥
810	s	γCH op Nap	⊥
842	s	γCH op Ph	⊥
883	m	βCC ip RC (breathing) + ν_{as} $C_{Ar}S$	||
1016	vw	βCC ip Ph	||
1073	w	βCC ip Nap + ν_{Al} CCC	||
1145	vw	βCH ip Nap	||
1175	w	βCH ip RC	||
1264	vw	βCH ip Nap	||
1463	vw	βCH_2 (scissoring) + βCC Nap	||
1487	s	βCC Nap + βCH_2 (scissoring)	||
1510	w	βCC Ph	||
1587	m	br Nap	||
1602	vs	br Ph	||
2223	vs	νCN	||

Key: ip—in-plane vibration; op—out-of-plane deformation; br—stretching and deformation vibrations of the ring (benzene ring); s—symmetrical; as—asymmetrical; Al—alkyl chain; Ar—aromatic ring; Ph—phenyl; Nap—naphthalene; RC—rigid core; ν—stretching; γ—deforming out of plane; β—deforming in plane; δ—deforming; vs—very strong; s—strong; m—medium; w—weak; vw—very weak; I_{rel}—relative intensity of the bands, $\nu\,[cm^{-1}]$—wavenumber.

Figure 4 shows a comparison of the simulated ($n = 7$) and experimental ($n = 3, 7$) spectra. We have previously shown, for similar N_{TB} dimers, that theoretical calculations of the IR spectra for an isolated molecule are better in N compared to N_{TB} experimental results; thus, the experimental results were captured from the N phase (440 K for both materials) [33]. The presented spectra were recorded as a function of incident radiation, where || represents the parallel absorbance component along the *z*-axis of the molecular system or the sample ordering axis, while ⊥ denotes the perpendicular absorbance component corresponding to the sample rubbing direction.

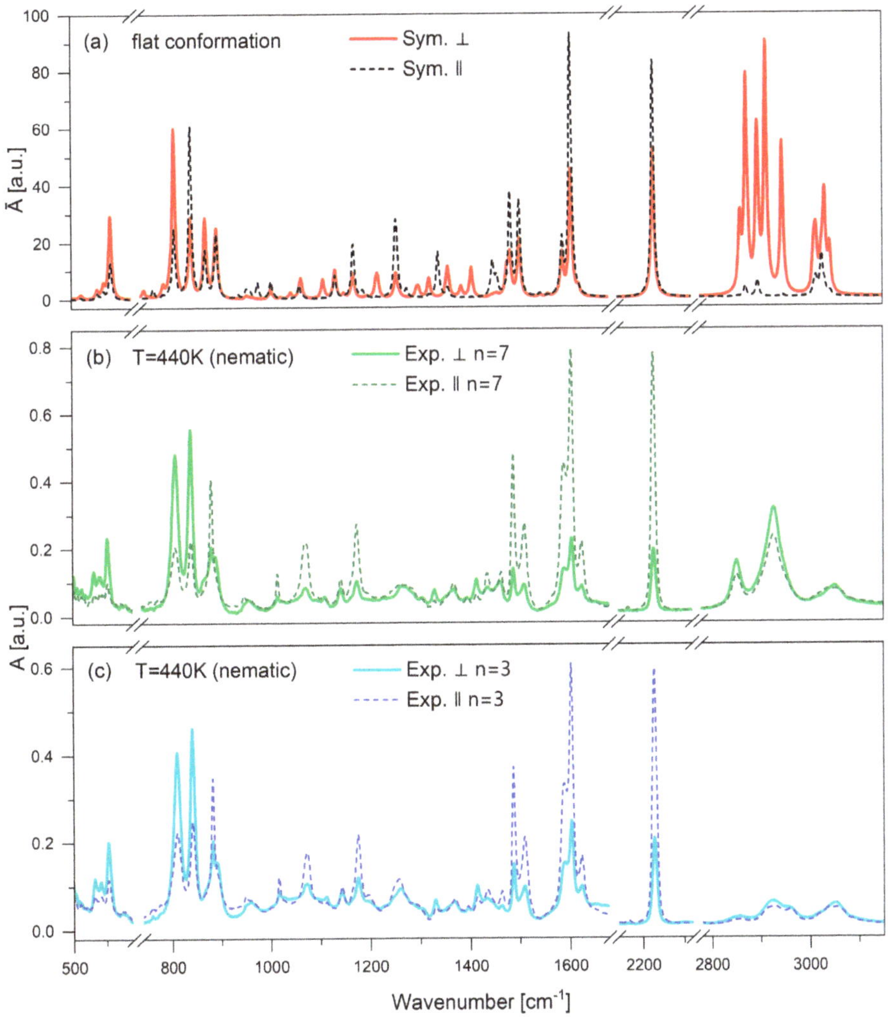

Figure 4. Comparison of the theoretical spectra (B3-LYP/6-311G (d,p)) with the polarized experimental spectra for the (CN)PNS*n*SNP(CN) (n = 3, 7) dimers in the region 500–3250 cm^{-1}. ||—parallel absorbance component (in the z-axis of the molecular system/sample ordering axis); ⊥—perpendicular absorbance component (sample rubbing direction). (**a**) Theoretical spectra of the (CN)PNS7SNP(CN) molecule, (**b**) experimental spectra of the (CN)PNS7SNP(CN) dimer in the N phase (440 K), and (**c**) experimental spectra of the (CN)PNS3SNP(CN) dimer in the N phase (440 K) (spectra are represented for 5 μm cell).

The experimental spectra for the dimers of different spacer lengths reveal a general uniformity for the majority of the vibrational bands. However, a prominent distinction is observed in the bands in 2800–3000 cm^{-1}, primarily associated with the stretching vibrations of the methylene (CH_2) groups. The shorter dimer displays reduced intensity in these bands, which can be attributed to the fewer CH_2 groups within its spacer. Compared to the theoretical spectra, both experiments show vibrational band overlap. Similarly, differences occurring within the spectral ranges characteristic of the vibrations of CH_2 groups in alkyl chains are also observed within the range of 1200–1300 cm^{-1}, corresponding to the γ_s and γ_{as} bending vibrations of CH_2.

Another noteworthy distinction between the two dimer lengths pertains to the dichroism of the band at 1600 cm^{-1}, defined as the ratio of the parallel to the perpendicular component. The dichroism of the shorter dimer ($R_3 = 2.44$) aligns better with the simulation ($R_{theor} = 2.06$), whereas the longer dimer exhibits reduced intensity in the perpendicular band with a dichroism of $R_7 = 3.44$. The bands at 1590 and 1600 cm^{-1} correspond to vibrations associated with mesogenic units. The former is primarily linked to naphthalene vibrations, while the latter involves benzene vibrations, with the transition dipole moment parallel to the long molecular axis. Compared to the simulation, there are notable differences in the distribution of intensity and absorbance. The higher-wavenumber shoulder appears more distinct, while the two lower bands coincide to a greater extent in the experimental results.

3.3. Simulated Spectra for a Hairpin Conformer

To assess the influence of extreme conformers on the experimental (CN)PNS7SNP(CN) spectrum in the high-temperature N phase (above 400 K), additional DFT calculations were conducted to determine the geometry and vibrational frequencies of more bent conformers. A single (CN)PNS7SNP(CN) molecule was arbitrarily bent into the so-called hairpin conformation, as shown in Figure 5, and optimized.

It is worth noting that in this case, the molecular reference system z-axis does not represent the long molecular axis, which, for the hairpin conformer, would be along the mesogen axis (see Figure 5). For this reason, the comparison of the polarized spectra for hairpin and bent conformers would be very difficult, since the distribution between the parallel and perpendicular transition dipole moment contributions for each of the bands is distorted (see Figure S1). It is more informative to compare unpolarized theoretical spectra for the bent and hairpin conformers vs. the experimental spectra in the N and N_{TB} phases (Figure 6).

The main differences between the two calculated conformers occurred in three areas: the range from 1050 to 1100 cm^{-1}, that from 1200 to 1375 cm^{-1}, and around 1450 cm^{-1}. Comparing the theoretical calculations with the experiment, one could conclude that the hairpin conformation bands which are not present in the bent conformation spectrum correlate with the experimental spectrum in the N phase. These hairpin bands are associated with the following linker vibrations:

- In-plane bending vibrations of the CH_2 groups (rocking δ_s^- CH_2) at 1076 cm^{-1};
- Out-of-plane bending vibrations of the CH_2 groups (wagging γ_s CH_2) at 1237 cm^{-1} and at 1367 cm^{-1};
- Out-of-plane bending vibrations of the CH_2 groups (twisting γ_{as} CH_2) at 1289 cm^{-1};
- In-plane bending vibrations of the CH_2 groups (scissoring β_s CH_2) at 1447 cm^{-1}.

There is only a significant contribution from linker vibration bands.

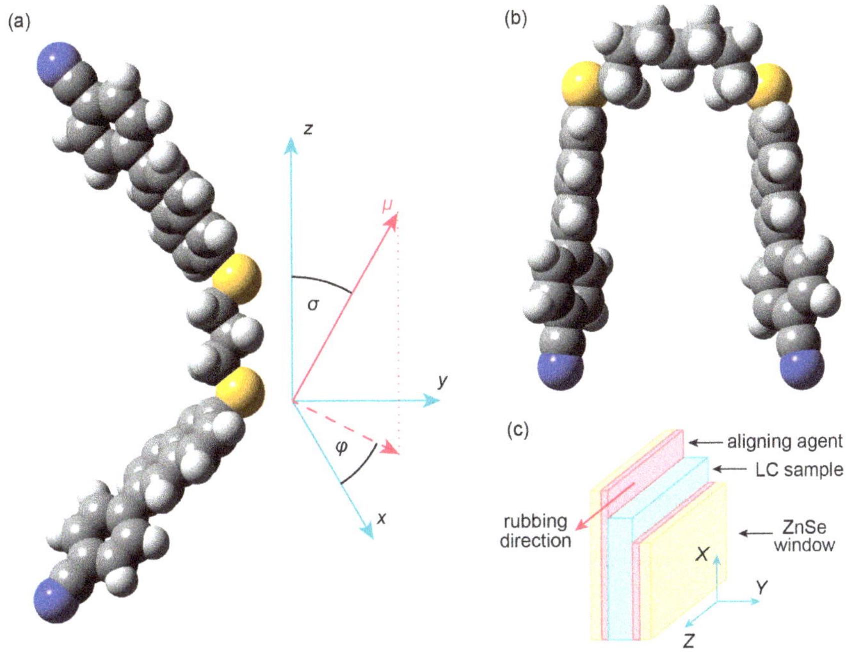

Figure 5. (**a**) Molecular frame of reference: z—long axis (bowstring); x—normal axis to the bent plane; y—bow arrow axis; σ—polar angle (between transition dipole μ and the z-axis of the molecule); φ is the azimuthal angle that the transition dipole makes with the x–z plane. (**b**) Hairpin conformer. (**c**) The orientation of the laboratory frame (X, Y, Z) for the planar sample. In the nematic phase, Z was an axis along and Y was an axis perpendicular to the optical axis (the optical axis coincided with the rubbing direction). In the N_{TB} phase, Z coincided with the helix axis, which was the symmetry element of the N_{TB} phase.

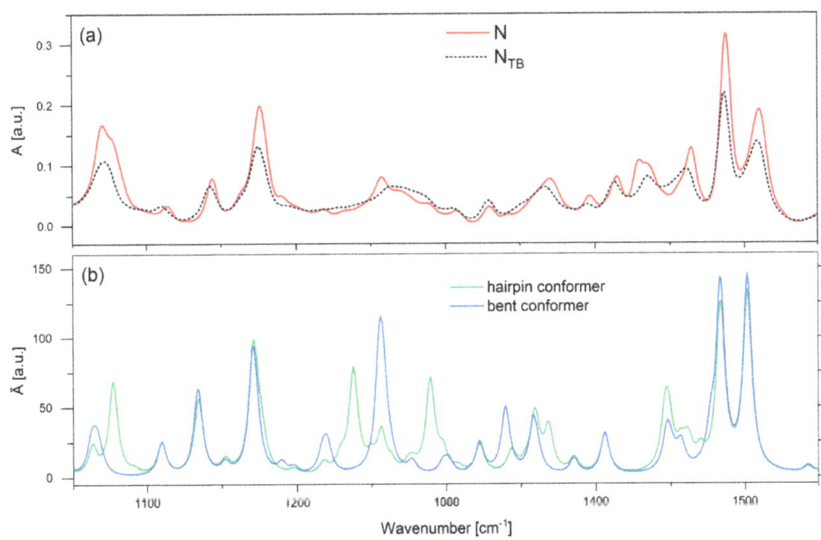

Figure 6. IR spectra of the (CN)PNS7SNP(CN) dimer in the wavenumber range of 1050–1550 cm^{-1}. (**a**) Experimental spectra: red solid line—the N phase; black dotted line—the N_{TB} phase. (**b**) Theoretical spectra (B3-LYP/6-311G (d,p)): green line—the hairpin; blue line—the bent conformers.

3.4. Experimental Results—Temperature Dependencies of Absorbance

We analyzed the temperature dependence of the average IR absorbance of the materials, defined as the average of each of the directional absorbance components (see Section 2.3), both in the N and N_{TB} phases. Within the N phase, the average absorbance increases with molecular density. However, below the N–N_{TB} transition, the behavior becomes dependent on the orientation of the transition dipole moment corresponding to each vibrational band. To track the reorganization of molecular order during the transition from the nematic to the twist–bend phase, we calculated the mean absorbance for several vibrational bands, each representing a distinct molecular segment. For longitudinal dipoles characterizing the overall molecular axis, we selected bands associated with the spacer (1463 cm^{-1}, CH_2 scissoring) and with naphthalene vibrations involving both the sulfide bridge and the spacer (1074, 1487, and 1586 cm^{-1}). Bands reflecting the behavior of the mesogenic arms (and thus the arms of the dimer) include 1603, 1510, and 2200 cm^{-1}, attributed to phenyl ring and cyano group vibrations. For these bands associated with the long dimer axis, a distinct decrease in average absorbance is observed at the onset of the N_{TB} phase. In contrast, the bands characterizing only the mesogenic arms show a delayed response—approximately 10 K below the transition temperature. Similar observations were made for the transverse dipole moment bands. Figure 7 presents the normalized mean IR absorbances for transversal and longitudinal transition dipole moments for both of the studied materials.

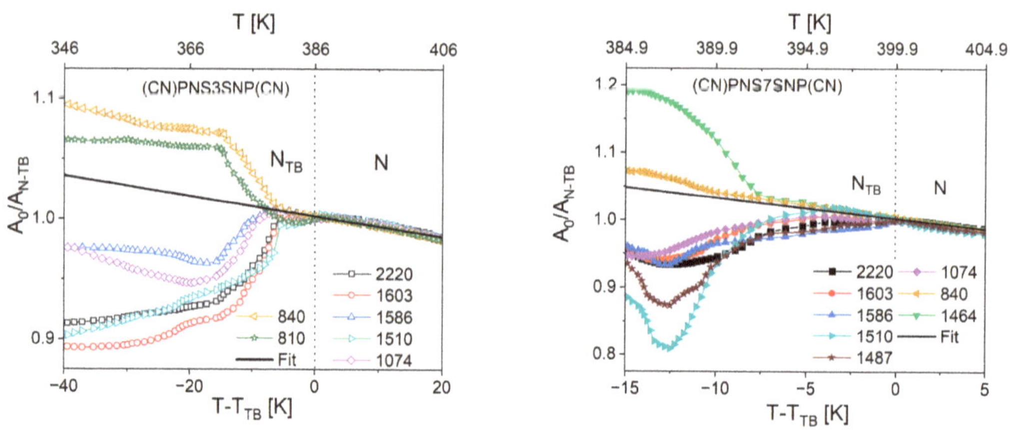

(**a**) (**b**)

Figure 7. Normalized IR absorbance vs. temperature behavior of the studied dimers: (**a**) for n = 3 (open symbols) and (**b**) for n = 7 (solid symbols). Longitudinal dipole moment bands: □/■—2200 cm^{-1} (νCN); ○/●—1603 cm^{-1} (br Ph); △/▲—1586 cm^{-1} (br Nap); ▷/► —1510 cm^{-1} (βCC ip Ph); ★—1487 cm^{-1} (βCC ip Nap +βCH_2 scissoring); ◇/◆—1074 cm^{-1} (βCC ip Nap + ν CCC). Transversal dipole moment bands: ◁/◀ —840 cm^{-1} (γCH op Ph); ☆ —810 cm^{-1} (γCH op Nap + C_{Ar}S); ▼—1464 cm^{-1} (βCH_2 scissoring + βCC Nap). Black line: nematic trend extrapolated to the lower temperature N_{TB} phase.

For transversal dipole moments, the average absorbance trends continue smoothly from the N phase into the N_{TB} phase. Only a slight kink is observed at the transition temperature for both dimers, while at approximately 10 K below the transition temperature, a sudden increase in the average absorbance is observed for all bands. This observation is generally consistent with previous reports on N_{TB}-forming dimers, such as those based on cyanobiphenyl cores, where an increase in mean absorbance was noted approximately 30 K below the N–N_{TB} transition [34]. Notably, in the earlier study, dimers with two thioether linkages exhibited a sharp increase at the transition, whereas the (CN)PNS*n*SNP(CN) dimers studied here show a more gradual change. Furthermore, in our case, no significant increase was observed in the lower-N_{TB}-temperature region. This is likely due to the rapid vitrification of the samples, which prevented the molecular reorganization necessary for the development of full transversal order, as molecular motion was significantly restricted.

In contrast, for longitudinal dipole moments, the average absorbance generally increases upon cooling through the nematic phase, although this trend slows as the system approaches the N–N_{TB} transition. At the transition point, a distinct drop in absorbance is observed, reflecting a reorganization of the molecular structure. At the same time, the transverse dipole moments continue to follow the nematic trend, showing a further increase in absorbance in the N_{TB} phase. These findings suggest that upon entering the twist–bend phase, longitudinal dipoles tend to adopt an antiparallel arrangement—manifested as a decrease in absorbance—while transverse dipoles align more cooperatively, leading to an increased absorbance. This indicates the onset of bond ordering, where specific orientations of neighboring dipoles become energetically favorable. Such behavior is not observed in the nematic phase, where the absorbance for both dipole types generally increases with cooling due to the rise in molecular density alone, without evidence of directional correlation.

Motivated by these observations, we sought to explore the molecular origin of this dipole reorganization. We hypothesized that the observed bond ordering may arise from short-range, lateral interactions between neighboring molecules, specific to the N_{TB} structure. To test this, we performed DFT simulations incorporating nearest-neighbor molecular arrangements, allowing us to assess how such local interactions can affect the transition dipole moments of individual vibrational bands.

4. Discussion

Focusing on the nearest-neighbor interactions and considering the high computational cost of DFT simulations for large molecular systems, we performed calculations for the (CN)PNS7 monomers. In the next step, six monomeric molecules, each in an optimized planar conformation, were arranged into a so-called sublayer (Figure 8). The molecules were placed parallel to each other, with an intermolecular separation of approximately 5 Å, and the cyano groups were arranged in an alternating fashion. This distance was estimated based on experimental data from resonant X-ray scattering (TReXS) studies of the N_{TB} phase for other cyanobiphenyl-based dimers [35]. However, no X-ray data were available for this specific material. The system was optimized using the B3LYP/6-311G (d,p) method.

Following optimization, the outermost molecules within the sublayer were frozen, and vibrational frequencies and intensities were calculated for the central molecule, which was surrounded by the neighboring species (see Figure 8; the selected molecule is marked in green). This methodology enabled the computation of the IR spectrum by determining the transition dipole moments for individual bands while incorporating the influence of the immediate molecular environment. The optimization process revealed nanoscale segregation within the pseudolayers—naphthalene cores preferentially aligned adjacent to one another, forming an aromatic domain, while the flexible spacer adopted a parallel configuration, contributing to the formation of an aliphatic layer. In this arrangement, the naphthalene units of adjacent molecules exhibited partial overlap with the phenyl groups, while the phenyl rings were oriented at an angle with respect to neighboring molecules. A partial overlap of the aromatic moieties and an attraction between the cyano groups and alkyl chains were also observed. The theoretical IR spectrum for the centrally positioned molecule, where intermolecular forces (IMFs) were explicitly considered, was compared with the spectrum of an isolated molecule as well as experimental spectra recorded in the N and N_{TB} phases (Figure 9).

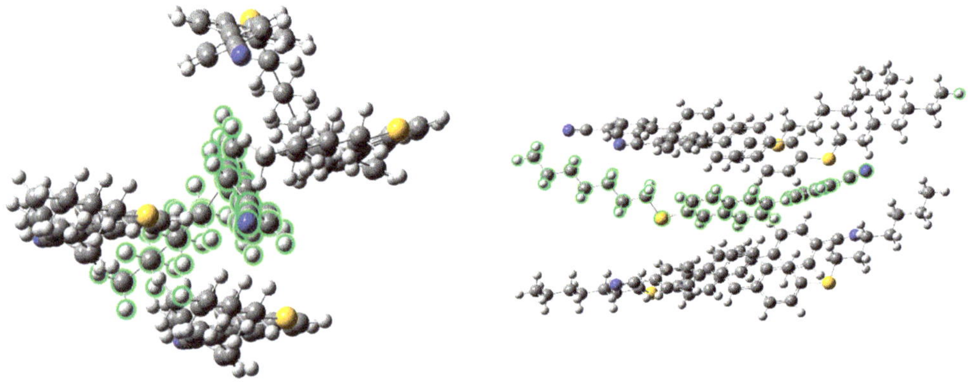

(a) (b)

Figure 8. View of the (CN)PNS7 monomer system. (**a**) A top view (X–Y plane) of the system with the arrangement of the five molecules after optimization by the B3LYP/6-311G (d,p) method. The distances between the molecules range from 5 to 6 Å. IR spectra were calculated for a molecule (highlighted in green) that is surrounded by other molecules. (**b**) A side view of the system.

A decrease in intensity was detected for the longitudinal dipole-related bands (wavenumbers: 880, 1060, 1490, 1510, 1600, and 2220 cm^{-1}) in the IMF-influenced system compared to an isolated monomer. This trend aligns with the experimental observations, where a decrease in mean absorbance for these bands was noted during the transition from the N to the N_{TB} phase (Figure 4). Conversely, an increase in intensity was noted for the transverse transition dipole moments, specifically at 807 and 850 cm^{-1}. This suggests a correlation of transverse dipoles perpendicular to the N order, commonly referred to as bond ordering. The intensity of the 807 cm^{-1} band increased by approximately 67% relative to the isolated molecule, while the 850 cm^{-1} band exhibited a 15% increase. The 807 cm^{-1} band corresponds to out-of-plane CH vibrations of the naphthalene moiety, whereas the 850 cm^{-1} band is associated with out-of-plane CH vibrations of the phenyl ring.

For the bands linked to longitudinal dipoles, a reduction in intensity was observed—approximately 57% for in-plane CC stretching modes of the naphthalene group (1590 cm^{-1}, 1060 cm^{-1}) and around 15% for characteristic vibrations of the phenyl group (1606 cm^{-1}, 1510 cm^{-1}). The smallest intensity change was noted for the CN stretching vibration at 2220 cm^{-1} (~10%). These computed values are consistent with experimental trends, where a decrease in the mean absorbance of longitudinal dipole bands (by ~30%) and an increase in transverse dipole bands (by ~25–35%) were recorded upon transitioning from the N to the N_{TB} phase.

It is important to emphasize that interactions involving a limited number of molecules do not fully account for the collective behavior of the system. To achieve a more comprehensive understanding, thermodynamic and statistical considerations—such as entropy and enthalpy contributions—must be incorporated. Therefore, the current calculations should be considered a preliminary step toward more advanced simulations involving the periodic density functional theory, which we are currently trying to carry out.

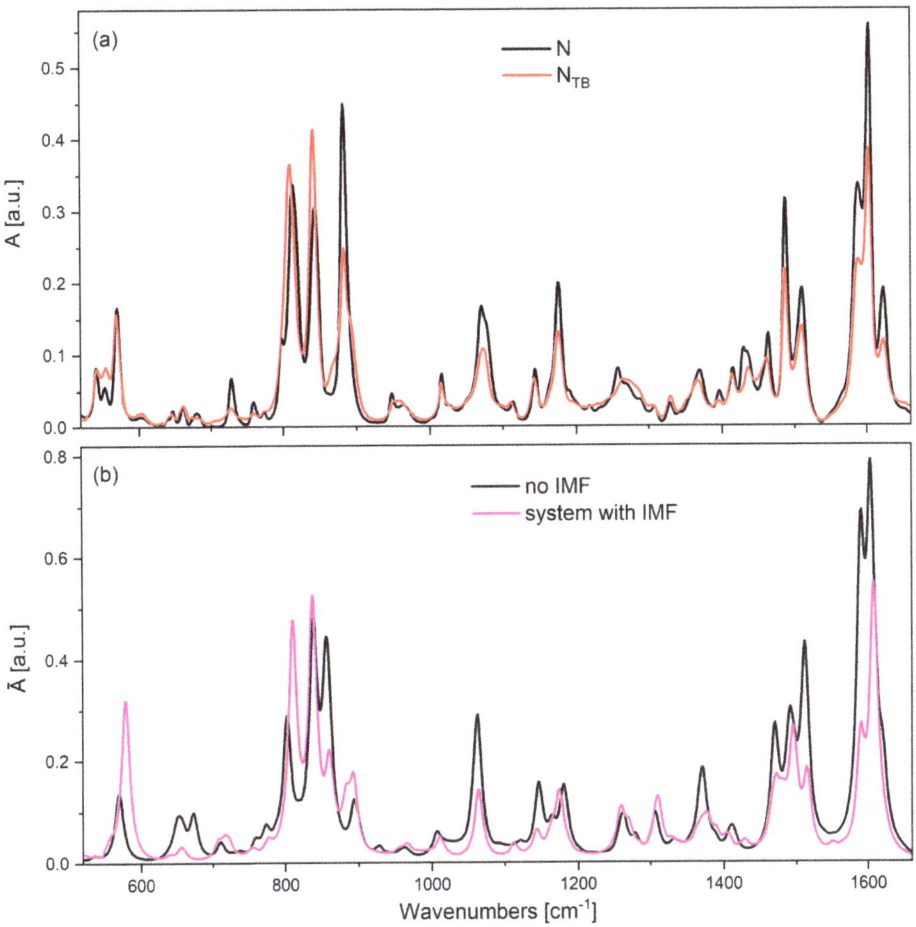

Figure 9. Comparison of the theoretical and experimental spectra for the (CN)PNS7SPN(CB) dimer. (**a**) Experimental spectra: solid black line—the N phase; solid red line—the N_{TB} phase (spectrum is represented for 5 mm cell). (**b**) Theoretical spectra for the (CN)PNS7 monomer: solid gray line—the isolated molecule (no IMF); solid magenta line—a system of five interacting molecules (with IMF).

5. Conclusions

It is widely accepted that the molecular curvature of dimers significantly influences the formation and stability of the N_{TB} phase. This study reinforces that view, demonstrating that precise molecular architecture, including both curvature and intermolecular interactions, is critical to realizing materials suitable for technological applications. Our combined theoretical and experimental investigation provides a comprehensive analysis of the structural and spectroscopic properties of naphthalene-based liquid-crystal dimers, offering insights into the mechanisms driving N_{TB} phase behavior.

Through detailed IR spectroscopic analysis, including both experimental and simulated spectra, we successfully assigned characteristic vibrational bands and identified distinct spectral features differentiating the N and N_{TB} phases. Importantly, we showed that the low torsional barrier of the sulfide bridge (~3 kJ/mol) allows neighboring molecules to induce significant conformational changes, influencing the vibrational dynamics of dimers in the N_{TB} phase. Our comparison between discrete low-energy conformers and the Boltzmann-averaged bending angle suggests that while flexibility exists, the average curvature remains relatively stable across conformational populations.

One of the most striking findings of this work is the evolution of dipole correlations during the N-to-N_{TB} phase transition. A pronounced decrease in longitudinal dipole absorbance, accompanied by an increase in transverse dipole absorbance, was observed. This shift indicates a reorganization of molecular alignment, where longitudinal dipoles become negatively correlated due to antiparallel mesogen arrangements, while transverse dipoles align cooperatively. These results point to a highly specific and cooperative reorientation process occurring at the molecular level during N_{TB} phase formation.

To elucidate the molecular basis of this self-organization, we performed DFT modeling incorporating nearest-neighbor interactions. These simulations underscore the crucial role of the local molecular environment in stabilizing the N_{TB} phase. In particular, weak noncovalent interactions—such as π–π stacking between aromatic cores and potential weak hydrogen bonds—emerge as important contributors to phase stability and heliconical ordering.

By integrating experimental IR spectroscopy and quantum chemical modeling, this study offers a multiscale understanding of how molecular structure and interactions govern N_{TB} phase formation in liquid-crystal dimers. These insights advance the fundamental understanding of self-organizing liquid crystalline systems and contribute to the rational design of new materials with tailored mesophases for future applications.

Supplementary Materials: The following supporting information can be downloaded at https://www.mdpi.com/article/10.3390/ma18091971/s1, Figure S1. Comparison of the theoretical spectra (B3-LYP/6-311G (d,p)) for the bent (top) and hairpin (bottom) conformers; Table S1. Theoretical and experimental frequencies, dichroism values, relative intensity, direction of the transition dipole moment, and approximate band assignments for (CN)PNS7SNP(CN).

Author Contributions: Conceptualization, K.M.; methodology, K.M. and B.L.; synthesis, Y.A.; formal analysis, K.M. and B.L.; investigation, B.L. and K.M.; writing—original draft, B.L. and K.M.; writing—review and editing, K.M., B.L. and Y.A.; visualization, B.L.; supervision, K.M. All authors have read and agreed to the published version of the manuscript.

Funding: This research was funded by the National Science Centre, Poland, grant no. 2020/39/O/ST5/03460.

Institutional Review Board Statement: Not applicable.

Informed Consent Statement: Not applicable.

Data Availability Statement: The original data presented in the study are openly available in RepOD at https://doi.org/10.18150/IUMWS0 [RepOD, V1]. Data available: raw IR measurement data and DFT simulation data. The date access: 25 April 2025.

Acknowledgments: K.M. and B.L. thank the National Science Centre for funding through project no. 2020/39/O/ST5/03460. We gratefully acknowledge Polish high-performance computing infrastructure PLGrid (HPC Center: ACK Cyfronet AGH) for providing computer facilities and support within computational grant no. PLG/2024/017005. K.M. would like to thank the Research Excellence Initiative (Edition III) of the University of Silesia in Katowice for allowing the purchase of the GRAMS/AI Spectroscopy Software Version 9, which has proven useful for the analysis of IR spectra.

Conflicts of Interest: The authors declare no conflicts of interest.

Abbreviations

The following abbreviations are used in this manuscript:

N	Conventional Nematic
N_{TB}	Twist–Bend Nematic

DFT	Density Functional Theory
DSC	Differential Scanning Calorimetry
IR	Infrared
POM	Polarizing Optical Microscopy
TReXS	Tender Resonant X-ray Scattering
IMFs	Intermolecular Forces
PN	Phenyl-naphthalene Ring Core
CN	Cyano Group
ZnSe	Zinc Selenide
FWHM	Full Width at Half Maximum
CB7CB	1″,7″-Bis(4-cyanobiphenyl-4′-yl) Heptane
(CN)PNS*n*SNP(CN)	6-(4-Cyanophenyl)-2-naphtalene-Based Dimers With the Heptyl and the Propyl Chain in the Spacer

References

1. Meyer, R.B. Structural Problems in Liquid Crystal Physics. In *Molecular Fluids*; Balian, R., Weil, G., Eds.; Les Houches Summer School in Theoretical Physics 1973; Gordon and Breach: New York, NY, USA, 1976; pp. 272–273.
2. Dozov, I. On the Spontaneous Symmetry Breaking in the Mesophases of Achiral Banana-Shaped Molecules. *Europhys. Lett.* **2001**, *56*, 247. [CrossRef]
3. Memmer, R. Liquid Crystal Phases of Achiral Banana-Shaped Molecules: A Computer Simulation Study. *Liq. Cryst.* **2002**, *29*, 483–496. [CrossRef]
4. Panov, V.P.; Nagaraj, M.; Vij, J.K.; Panarin, Y.P.; Kohlmeier, A.; Tamba, M.G.; Lewis, R.A.; Mehl, G.H. Spontaneous Periodic Deformations in Nonchiral Planar-Aligned Bimesogens with a Nematic-Nematic Transition and a Negative Elastic Constant. *Phys. Rev. Lett.* **2010**, *105*, 167801. [CrossRef] [PubMed]
5. Cestari, M.; Diez-Berart, S.; Dunmur, D.A.; Ferrarini, A.; de la Fuente, M.R.; Jackson, D.J.B.; Lopez, D.O.; Luckhurst, G.R.; Perez-Jubindo, M.A.; Richardson, R.M.; et al. Phase Behavior and Properties of the Liquid-Crystal Dimer 1″,7″-Bis(4-Cyanobiphenyl-4′-Yl) Heptane: A Twist-Bend Nematic Liquid Crystal. *Phys. Rev. E* **2011**, *84*, 031704. [CrossRef] [PubMed]
6. Borshch, V.; Kim, Y.-K.; Xiang, J.; Gao, M.; Jákli, A.; Panov, V.P.; Vij, J.K.; Imrie, C.T.; Tamba, M.G.; Mehl, G.H.; et al. Nematic Twist-Bend Phase with Nanoscale Modulation of Molecular Orientation. *Nat. Commun.* **2013**, *4*, 2635. [CrossRef]
7. Zhu, C.; Tuchband, M.R.; Young, A.; Shuai, M.; Scarbrough, A.; Walba, D.M.; Maclennan, J.E.; Wang, C.; Hexemer, A.; Clark, N.A. Resonant Carbon K-Edge Soft X-Ray Scattering from Lattice-Free Heliconical Molecular Ordering: Soft Dilative Elasticity of the Twist-Bend Liquid Crystal Phase. *Phys. Rev. Lett.* **2016**, *116*, 147803. [CrossRef]
8. Chen, D.; Porada, J.H.; Hooper, J.B.; Klittnick, A.; Shen, Y.; Tuchband, M.R.; Korblova, E.; Bedrov, D.; Walba, D.M.; Glaser, M.A.; et al. Chiral Heliconical Ground State of Nanoscale Pitch in a Nematic Liquid Crystal of Achiral Molecular Dimers. *Proc. Natl. Acad. Sci. USA* **2013**, *110*, 15931–15936. [CrossRef]
9. Panov, V.P.; Balachandran, R.; Nagaraj, M.; Vij, J.K.; Tamba, M.G.; Kohlmeier, A.; Mehl, G.H. Microsecond Linear Optical Response in the Unusual Nematic Phase of Achiral Bimesogens. *Appl. Phys. Lett.* **2011**, *99*, 261903. [CrossRef]
10. Xiang, J.; Li, Y.; Li, Q.; Paterson, D.A.; Storey, J.M.D.; Imrie, C.T.; Lavrentovich, O.D. Electrically Tunable Selective Reflection of Light from Ultraviolet to Visible and Infrared by Heliconical Cholesterics. *Adv. Mater.* **2015**, *27*, 3014–3018. [CrossRef]
11. Mrukiewicz, M.; Iadlovska, O.S.; Babakhanova, G.; Siemianowski, S.; Shiyanovskii, S.V.; Lavrentovich, O.D. Wide Temperature Range of an Electrically Tunable Selective Reflection of Light by Oblique Helicoidal Cholesteric. *Liq. Cryst.* **2019**, *46*, 1544–1550. [CrossRef]
12. Vanakaras, A.G.; Samulski, E.T.; Photinos, D.J. Entropy Stabilized Form Chirality in Curved Rod Nematics: Structure and Symmetries. *Soft Matter* **2025**, *21*, 1341–1352. [CrossRef]
13. Vanakaras, A.G.; Photinos, D.J. A Molecular Theory of Nematic–Nematic Phase Transitions in Mesogenic Dimers. *Soft Matter* **2016**, *12*, 2208–2220. [CrossRef] [PubMed]
14. Vanakaras, A.G.; Photinos, D.J. Molecular Dynamics Simulations of Nematic Phases Formed by Cyano-Biphenyl Dimers. *Liq. Cryst.* **2018**, *45*, 2184–2196. [CrossRef]
15. Samulski, E.T.; Vanakaras, A.G.; Photinos, D.J. The Twist Bend Nematic: A Case of Mistaken Identity. *Liq. Cryst.* **2020**, *47*, 2092–2097. [CrossRef]
16. Samulski, E.T. The Ever Elusive, Yet-to-Be-Discovered Twist-Bend Nematic Phase. *Crystals* **2023**, *13*, 1648. [CrossRef]
17. Archbold, C.T.; Mandle, R.J.; Andrews, J.L.; Cowling, S.J.; Goodby, J.W. Conformational Landscapes of Bimesogenic Compounds and Their Implications for the Formation of Modulated Nematic Phases. *Liq. Cryst.* **2017**, *44*, 2079–2088. [CrossRef]

18. Mandle, R.J.; Goodby, J.W. Molecular Flexibility and Bend in Semi-Rigid Liquid Crystals: Implications for the Heliconical Nematic Ground State. *Chem. Eur. J.* **2019**, *25*, 14454–14459. [CrossRef]
19. Pocock, E.E.; Mandle, R.J.; Goodby, J.W. Experimental and Computational Study of a Liquid Crystalline Dimesogen Exhibiting Nematic, Twist-Bend Nematic, Intercalated Smectic, and Soft Crystalline Mesophases. *Molecules* **2021**, *26*, 532. [CrossRef]
20. Kumar, A. Dependency of the Twist-Bend Nematic Phase Formation on the Molecular Shape of Liquid Crystal Dimers: A View through the Lens of DFT. *J. Mol. Liq.* **2022**, *354*, 118858. [CrossRef]
21. Górecka, E.; Salamończyk, M.; Zep, A.; Pociecha, D.; Welch, C.; Ahmed, Z.; Mehl, G.H. Do the Short Helices Exist in the Nematic TB Phase? *Liq. Cryst.* **2015**, *42*, 1–7. [CrossRef]
22. Mandle, R.J. The Dependency of Twist-Bend Nematic Liquid Crystals on Molecular Structure: A Progression from Dimers to Trimers, Oligomers and Polymers. *Soft Matter* **2016**, *12*, 7883–7901. [CrossRef]
23. Arakawa, Y.; Komatsu, K.; Tsuji, H. Twist-Bend Nematic Liquid Crystals Based on Thioether Linkage. *New J. Chem.* **2019**, *43*, 6786–6793. [CrossRef]
24. Cruickshank, E.; Salamończyk, M.; Pociecha, D.; Strachan, G.J.; Storey, J.M.D.; Wang, C.; Feng, J.; Zhu, C.; Gorecka, E.; Imrie, C.T. Sulfur-Linked Cyanobiphenyl-Based Liquid Crystal Dimers and the Twist-Bend Nematic Phase. *Liq. Cryst.* **2019**, *46*, 1595–1609. [CrossRef]
25. Merkel, K.; Kocot, A.; Welch, C.; Mehl, G.H. Soft Modes of the Dielectric Response in the Twist–Bend Nematic Phase and Identification of the Transition to a Nematic Splay Bend Phase in the CBC7CB Dimer. *Phys. Chem. Chem. Phys.* **2019**, *21*, 22839–22848. [CrossRef] [PubMed]
26. Meyer, C.; Blanc, C.; Luckhurst, G.R.; Davidson, P.; Dozov, I. Biaxiality-Driven Twist-Bend to Splay-Bend Nematic Phase Transition Induced by an Electric Field. *Sci. Adv.* **2020**, *6*, eabb8212. [CrossRef]
27. Knežević, A.; Sapunar, M.; Buljan, A.; Dokli, I.; Hameršak, Z.; Kontrec, D.; Lesac, A. Fine-Tuning the Effect of π–π Interactions on the Stability of the NTB Phase. *Soft Matter* **2018**, *14*, 8466–8474. [CrossRef]
28. Arakawa, Y.; Ishida, Y.; Tsuji, H. Ether- and Thioether-Linked Naphthalene-Based Liquid-Crystal Dimers: Influence of Chalcogen Linkage and Mesogenic-Arm Symmetry on the Incidence and Stability of the Twist-Bend Nematic Phase. *Chem. Eur. J.* **2020**, *26*, 3767–3775. [CrossRef]
29. Frisch, M.J.; Trucks, G.W.; Schlegel, H.B.; Scuseria, G.E.; Robb, M.A.; Cheeseman, J.R.; Scalmani, G.; Barone, V.; Petersson, G.A.; Nakatsuji, H.; et al. *Gaussian 16, Revision C.01*; Gaussian, Inc.: Wallingford, CT, USA, 2016.
30. Becke, A.D. Density-Functional Exchange-Energy Approximation with Correct Asymptotic Behavior. *Phys. Rev. A* **1988**, *38*, 3098–3100. [CrossRef]
31. Hertwig, R.H.; Koch, W. On the Parameterization of the Local Correlation Functional. What Is Becke-3-LYP? *Chem. Phys. Lett.* **1997**, *268*, 345–351. [CrossRef]
32. Dennington, R.; Keith, T.A.; Millam, J.M. *GaussView*, Version 6.1; Semichem Inc.: Shawnee Mission, KS, USA, 2016.
33. Merkel, K.; Loska, B.; Arakawa, Y.; Mehl, G.H.; Karcz, J.; Kocot, A. How Do Intermolecular Interactions Evolve at the Nematic to Twist–Bent Phase Transition? *Int. J. Mol. Sci.* **2022**, *23*, 11018. [CrossRef]
34. Kocot, A.; Loska, B.; Arakawa, Y.; Merkel, K. Structure of the Twist-Bend Nematic Phase with Respect to the Orientational Molecular Order of the Thioether-Linked Dimers. *Phys. Rev. E* **2022**, *105*, 044701. [CrossRef] [PubMed]
35. Cao, Y.; Feng, J.; Nallapaneni, A.; Arakawa, Y.; Zhao, K.; Zhang, H.; Mehl, G.H.; Zhu, C.; Liu, F. Deciphering Helix Assembly in the Heliconical Nematic Phase via Tender Resonant X-Ray Scattering. *J. Mater. Chem. C* **2021**, *9*, 10020–10028. [CrossRef]

materials

MDPI

Article

Increase in Auxeticity Due to the Presence of a Disordered Crystalline Phase of Hard Dumbbells Within the Nanolayer–Nanochannel Inclusion Introduced to the f.c.c. Hard Sphere Crystal

Jakub W. Narojczyk

Institute of Molecular Physics, Polish Academy of Sciences, M. Smoluchowskiego 17, 60-179 Poznań, Poland; narojczyk@ifmpan.poznan.pl

Abstract: To obtain materials or metamaterials with desired elastic properties that are tailor-made for a particular application, it is necessary to design a new material or composite (which may be cumbersome) or to modify the structure of existing materials in order to change their properties in the desired direction. The latter approach, although also not easy, seems favourable with respect to parameters like costs and time-to-market. Despite the fact that elastic properties are one of the oldest studied physical parameters of matter, our understanding of the processes at the microstructural level, that are behind these properties, is still far from being complete. The present work, with the help of Monte Carlo computer simulations, aims to broaden this knowledge. The previously studied model crystal of hard spheres, containing a combined nanolayer and nanochannel inclusions, is revisited. This periodic model crystal has been extended to include a degree of disorder in the form of degenerate crystalline phase by introducing a degenerate crystalline phase within its structure. The inclusion has been transformed (without changes to its shape, size, or orientation) by randomly connecting the neighbouring spheres into di-atomic molecules (dumbbells). The impact of this modification on elastic properties has been investigated with the help of the Parrinello–Rahman approach in the isothermal–isobaric ensemble (NpT). It has been shown, that the presence of the degenerate crystalline phase of hard dumbbells in the system leads to a significant decrease in the Poisson's ratio in [110]-direction ($\nu = -0.235$) and an overall enhancement of the auxetic properties.

Keywords: auxetics; negative Poisson's ratio; nanolayers; hard spheres; inclusions; Monte Carlo simulations

1. Introduction

The elastic properties are one of the earliest studied characteristics of materials. To date, most of these properties are very well measured and documented for most materials that are used. This, however, does not help us when facing a problem that requires a material with properties that go beyond what can be achieved for conventional ones. The demand for materials with new or specific properties is proportional to the increasing number of possible applications. It was not long ago when the materials exhibiting negative Poisson's ratio [1] (although theoretically possible) were thought to be non-existent. This was due to the fact that the ability of some natural materials to exhibit negative Poisson's ratio in only some of the directions, e.g., most of the cubic crystalls [2], has not yet been noticed. At present, *auxetics* [3], as they came to be called, are a well established class of materials that are undergoing constant development [4] by scientists and engineers. It has been almost 40 years since the pioneering works by Lakes [5] (who fabricated the first man made auxetic foam) and Wojciechowski [6–8], who formulated and solved the first thermodynamically stable model showing and explaining the nature of these extraordinary properties. Despite the significant effort in the study of auxetics, their practical applications, to date, are rare. However, a positive sign is their increase in recent years. The most noteworthy are the attempts to apply auxetic materials and structures in medical, bioengineering, and personal protection areas, as they have the chance to directly impact human health and wellbeing. In particular, the creation of auxetic micropatterned cardiac patches, obtained by excimer laser microablation, with the mechanical properties of the patches matching those of the native heart of the patients' tissue, were reported by Stevens et al. [9]. A recent report by Kaufman et al. [10] presented an animal study, where modified stents that incorporated an auxetic structure were implanted to adult ovine. Their findings have shown that auxetic stents immediately increased the vein inflow area. Graziosi et al. investigated the feasibility of using auxetic patterns to enhance kinesiology tapes [11] used in the treatment of musculoskeletal injuries that are frequent among young athletes and physically active adults. Sport is another area where the auxetic materials are utilized in the form of enhanced personal protection apparel end equipment [12,13]. The design and fabrication of such personal apparel will also require enhanced fabrics. A recent review by Chang and Hu [14] discussed auxetic knitted, woven, and non-woven fabrics. Applications of auxetics in engineering sectors can also be found, e.g., in furniture design, where the use of auxetic springs [15] and dowels [16] has been investigated. The auxetic materials have been studied both theoretically and experimentally [17–21]. The latter focused on the development and fabrication of (among others) negative Poisson's ratio polymers [22,23], composites [24,25], foams [26–28], fabrics [29–32], and metamaterials [33]. Alderson et al. first showed that it is possible to produce auxetic polypropylene fibres on an industrial, large-scale extruder [22] and produced fibres with a high degree of auxeticity. Novak et al. fabricated a chiral auxetic sandwich panel composite [24] and performed ballistic testing, both experimentally and numerically, to investigate the impact resistance of such a structure. Dalcanale et al. [25] proposed a promising solution to the problem of generating artificial molecular auxetics. The proposed auxetic polymer did not mimic metamaterials at the molecular level. Rather, it relied on the approach of mechanically driven conformational expansion. The matrix of the proposed polymer of intrinsic microporosity presents a rigid structure that is capable of transferring the mechanical stress into a conformational expansion of the crosslinking units. On the other hand, the theoretical studies on auxetic materials focus mainly on the development of new models that show a negative Poisson's ratio [34–48], which are later solved with a variety of computer simulation techniques, from which Finite Element Method [42,49–52] for macroscopic models and Monte Carlo [53–56] or molecular dynamics [57] for atomistic models, are the two most popular approaches. The development of auxetic materials can be roughly divided into two categories: the design of new materials or models with auxetic properties, and the study of available materials or models in order to enhance their elastic properties. The atomic models are especially interesting as they relatively easily allow testing various modifications to materials' structures and assess their impact on macroscopic elastic properties. In natural volumetric systems, such as liquid crystals, characterized by a regular periodic arrangement of atoms or molecules that exhibit isotropic or anisotropic properties dependent on the alignment, the inclusions and/or defects, can significantly affect there elasticity and viscosity [58,59]. It has been shown that modifying the crystalline structure using inclusions to introduce different types of particles into a material, one can significantly change its elastic properties [60,61]. However, depending on the size, shape, or orientation of the inclusions, the result may be either enhancement [60] or removal [61] of auxetic properties. This shows that we need to gain a deeper understanding on how the microstructural modifications impact macroscopic elastic properties (e.g., Poisson's ratio).

In this work, the impact of the disorder introduced into an otherwise periodic crystalline system will affect its Poisson's ratio. Disorder is an important aspect as, in reality, we always deal with a certain level of defects or particle size dispersion [54,62–64], and their impact on the elastic properties has to be studied and taken into consideration in practical applications. Disorder may be considered in a variety of aspects. In 1987, Wojciechowski theoretically predicted [65] the existence of an aperiodic solid phase in the two-dimensional (2D) system of hard dumbbells when they are close to dimers, i.e., when the distance between the centres of the discs forming a dumbbell is close to the diameter of the discs. This phase was initially coined by him as the disordered crystal [65] and, finally, the *degenerate crystal* (DC) [66–68]. That prediction was soon confirmed by Monte Carlo simulations of systems of 2D dimers [66,67,69] and dumbbells [70]. The existence of the DC phase was also confirmed for three-dimensional dimers [71]. The characteristic feature of a degenerate crystal is the lack of periodicity in positions and orientations of molecules, where, at the same time, the positions of atoms that form the molecules are periodic and can form, e.g., a hexagonal 2D lattice or f.c.c. and h.c.p. in 3D. The use of this phase was chosen for this study because it can be easily applied to existing models with studied previously inclusions [61] and its impact on elastic properties when used with additional complex inclusion sets have not been investigated. As mentioned in the previous paragraph, the introduction of combined nanolayer and nanochannel inclusions into the hard sphere f.c.c. crystal resulted in an increase in Poisson's ratio values and the removal of the auxetic properties. The objective of this work is to introduce a disordered crystalline phase into such a model and to investigate its impact on the elastic properties of such system.

This article has the following structure. In Section 2, the studied models are described. The research method is briefly reviewed in Section 3. The discussion of the results is conducted in Section 4. Section 5 contains the summary and conclusions.

2. The Model

The model studied in this work is an extension of the model studied in [61]. It consists of N hard spheres with a diameter σ, initially forming the f.c.c. lattice at close packing. The value σ is also treated as the unit of length. For such a system, inclusions are introduced by replacing some of the particles with other hard spheres of a different diameter (σ'). The choice of the shape and size of the inclusions is arbitrary; here, only the combination of a single nanochannel and nanolayer is considered. The nanochannel is oriented in the [001]-direction and has a radius equal to σ. All spheres forming the original system (which will be further referenced as the *matrix crystal*), whose centres of mass are positioned at close packing within this distance from a selected crystalline axis, are replaced with the σ'-spheres. The nanochannel radius has been selected, such that only the particles lying directly on the crystalline axis and their nearest neighbours are included. The nanolayer is oriented orthogonally to the nanochannel, and, similarly, all spheres that form a selected crystallographic plane are replaced with inclusion spheres. This is illustrated in Figure 1. Such a case, where the inclusion is formed by hard spheres, has been studied in [61]. Here, an additional condition (constraint) is introduced into the model by randomly paring the neighbouring inclusion spheres into diatomic molecules—hard dumbbells. Thus, the degenerate crystal phase is obtained within the inclusion. The elastic properties of a 3D degenerate crystalline system with nanochannel inclusions filled with hard spheres has been studied in [72].

The interatomic interactions in this work are modelled with the hard sphere (HS) potential of the form:

$$\beta u_{ij} = \begin{cases} \infty, & r_{ij} < \sigma_{ij}, \\ 0, & r_{ij} \geq \sigma_{ij}, \end{cases} \tag{1}$$

where r_{ij} is the distance between the centres of spheres i and j and σ_{ij} stand for the sum of the radii of these spheres, $\beta = 1/(k_B T)$, k_B is the Boltzmann constant, and T is the temperature. This simple, athermal and purely geometric interaction is a fundamental reference model in the theory of liquids and condensed matter physics [73,74], and is also extremely useful in the study of elastic properties, as hard models can exhibit a negative Poisson's ratio. As mentioned, this geometrical interaction is only sensitive to the change in dimensions and shape of interacting particles. Thus, the inclusion spheres have the diameters $\sigma' \neq \sigma$. The impact of the inclusions will be analyzed with respect to the ratio σ'/σ, which is the ratio of the inclusion to the matrix sphere diameters.

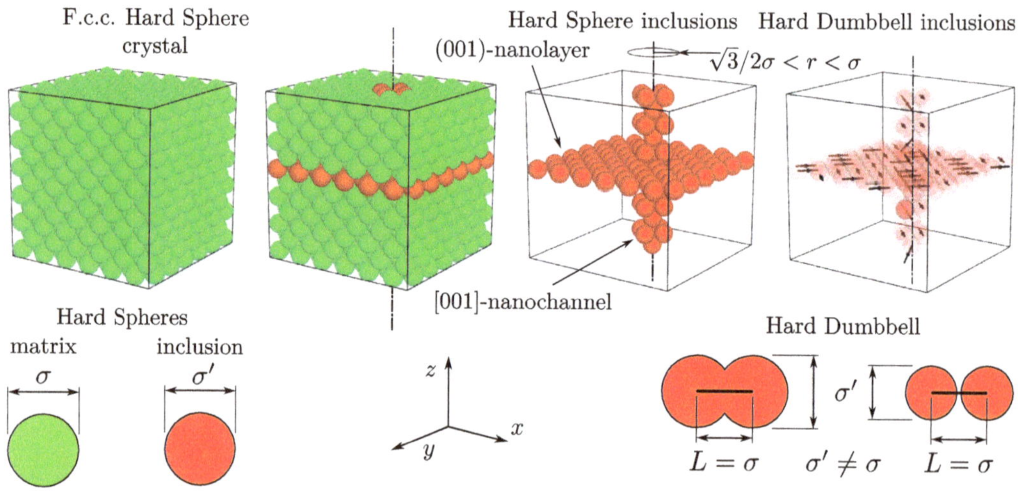

Figure 1. Visualization of the studied models. The f.c.c. HS crystal with the [001]-nanochannel and (001)-nanolayer inclusions filled by either hard spheres or hard dumbbells. In all cases, inclusion spheres have the diameter σ', that is different from the diameter of matrix spheres (σ).

It is important to note that the described model is considered in the periodic boundary conditions. The studied supercell, formed by selected N particles, is surrounded with the 26 identical copies of itself. Thus, one obtains a system with periodically stacked nanolayers combined with a periodic array of nanochannels. Visualization of periodic boundaries is given in [61] (see Figure 1 therein).

3. The Method

3.1. Elastic Properties

The Parrinello–Rahman [75–77] idea has been used with the Monte Carlo simulation method to determine the elastic properties of the model described in the previous section. Similar to [61], modelled supercells have been studied in the isobaric–isothermal ensemble (NpT), i.e., the number of particles, external hydrostatic pressure and temperature were kept constant during the simulation. To briefly summarize, the shape fluctuations [75–77] of the parallelepiped volume that contained the studied system, and described by a symmetric matrix **h** (formed by vectors defining its edges), allow for determining the elastic strain tensor ε:

$$\varepsilon = \frac{1}{2}\left(\mathbf{h}_{\mathrm{p}}^{-1}.\mathbf{h}.\mathbf{h}.\mathbf{h}_{\mathrm{p}}^{-1} - \mathbf{I}\right), \tag{2}$$

where **I** is a unit matrix. The reference matrix $\mathbf{h}_{\mathrm{p}}$ is equivalent to $\langle \mathbf{h} \rangle$, i.e., the average value of the **h** matrix under the dimensionless pressure $p^* = p\beta\sigma^3$, where $\beta = 1/k_B/T$, T is the temperature, and k_B is the Boltzmann constant. The strain tensor elements can be used to calculate the complete elastic compliance tensor S [61,77], a fourth rank symmetric tensor of 81 components in the case of the lowest symmetry. The knowledge of S allows one to fully describe the elastic properties of any crystal. It can be used to calculate the Poisson's ratio in any crystallographic direction using the form [78,79]:

$$\nu_{nm} = -\frac{m_\alpha m_\beta S_{\alpha\beta\gamma\delta} n_\gamma n_\delta}{n_\zeta n_\eta S_{\zeta\eta\kappa\lambda} n_\kappa n_\lambda}. \tag{3}$$

$\vec{n}$ and $\vec{m}$ are the pair of unit vectors in a mutually orthogonal direction in which, respectively, the external load is applied to the system and in which Poisson's ratio is measured. In Equation (3), indices n and m correspond to these vectors, where n_α and m_β are their respective components. It should be noted that the Einstein summation convention has been used on Greek indices. Moreover, for clarity, the S tensor has been replaced with the elastic compliance matrix **S** (a symmetric square matrix of dimension six) using the Voigt representation [80]. The Latin indices for S_{ij} elements are equal to $i, j = 1, \ldots, 6$. For further details on the method used here to calculate elastic properties of the studied models, please see [61,77].

3.2. Computation Details

The selected size of the sample used for simulations was $N = 864$ spheres, which corresponds to $6 \times 6 \times 6$ f.c.c. unit cells. The combined nanochannel and nanolayer inclusions consisted of $N_{\text{inc}} = 101$ (11.68% of N). The selected size has been previously verified to provide accurate results, by simulating systems doubled in all three directions [81]. To avoid the diffusion of particles from the inclusion to the matrix and to facilitate a comparison of the results with the ones reported in [61], the external hydrostatic (dimensionless) pressure has been set to $p^* = 100$. It is worth adding that this is only one order of magnitude higher than the pressure under which the hard sphere system undergoes melting. Different values of pressure have also been studied, and they will be reported in a separate article. As discussed in Section 2, the changes in the particle diameters will drive changes in the elastic properties of the system. Thus, simulations have been performed with σ'/σ between 0.95 and 1.05. This corresponds to simulations done in [61] and is within the range of values for which the stable tetragonal systems with molecular inclusions are obtained. The Monte Carlo simulations are 10^7 MC cycles long. The MC cycle is understood as a period in which all particles in the system undergo one trial translation (and rotation for molecules). The tests for the volume changes have been set to occur at every $\sqrt{N}$'th trial move of the particles. The first 10% of MC cycles have been treated as the period in which the system reaches the state of thermodynamic equilibrium and, thus, have been discarded from calculations. For every tested value of σ'/σ, at least 20 independent simulations have been performed for systems with spherical inclusions [61] and 200 simulations for systems with molecular inclusions. In the latter case, the molecular position and orientation has been randomly generated to better sample the degenerate crystal system.

4. Results and Discussion

The inclusion introduced to the f.c.c. crystal, whether composed of spheres or with the disordered phase of diatomic dumbbells, results in a shape change in the sample. In Figure 2, one can observe how the unit supercell changes from cubic ($\sigma'/\sigma = 1$) to cuboid with a square base. The x and y dimensions increase to compensate for the increasing atomic diameters in the nanolayer, and the z direction shrinks as the remaining particles have more space to compress under pressure. Figure 2b shows the off-diagonal components of the box matrix in relation to h_{11}, and it can be seen that they are at least three orders of magnitude less than the diagonal ones. This shows that the supercell has the cuboid shape. Following the change in shape, the symmetry of the system also changes due to the presence of inclusion. In Figure 3, the components of the elastic compliance matrix (**S**) have been presented and compared with the values of the f.c.c. cubic crystal ($\sigma'/\sigma = 1$). One can see that lowering the σ'/σ value (decreasing atomic diameters of molecules) does not impact elastic properties significantly. However, when $\sigma'/\sigma > 1$, the change in crystalline symmetry is clear. It can be noted that the following relations $S_{11} = S_{22}$, $S_{44} = S_{55}$, and $S_{13} = S_{23}$ are fulfilled, and that the symmetry of the crystal with such inclusion is tetragonal (422 symmetry class [80]). Thus, the elastic compliance matrix takes the following form:

$$\mathbf{S} = \begin{bmatrix} S_{11} & S_{12} & S_{13} & 0 & 0 & 0 \\ \cdot & S_{11} & S_{13} & 0 & 0 & 0 \\ \cdot & \cdot & S_{33} & 0 & 0 & 0 \\ \cdot & \cdot & \cdot & S_{44} & 0 & 0 \\ \cdot & \cdot & \cdot & \cdot & S_{44} & 0 \\ \cdot & \cdot & \cdot & \cdot & \cdot & S_{66} \end{bmatrix}. \tag{4}$$

The same symmetry has been observed in the case of the same combination of inclusions filled only by spheres [61]. An important remark should be made here, that as a common reference point to both systems with inclusions, the pure f.c.c. crystal of hard spheres with the diameters σ has been selected. One should be aware that the presence of molecular bonds from dumbbell particles at $\sigma'/\sigma = 1$ (in that case they are called dimers) will also have a certain (small) impact on the elastic properties. This difference is very small compared with the effects discussed in this work (one can analyse the changes in Figure 3 between $\sigma'/\sigma = 1$ and 0.99), and thus is ignored. In Figure 4, the comparison of the non-zero **S** matrix elements between atomic and molecular inclusions have been presented. It can be seen that additional constraints in the form of disordered molecular bonds exert more significant changes in the elastic properties than in the case of atomic systems (especially for S_{44}, S_{66}(b) and S_{12}, S_{13}(c)).

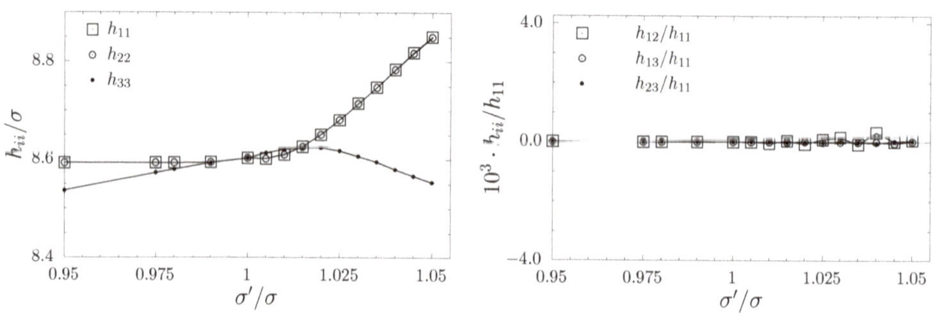

(**a**) (**b**)

Figure 2. The diagonal elements of the box matrix h_{ii} (**a**) and the ratio of the off-diagonal box matrix elements and h_{11} (**b**) for systems with dumbbell inclusions at $p^* = 100$, plotted against σ'/σ.

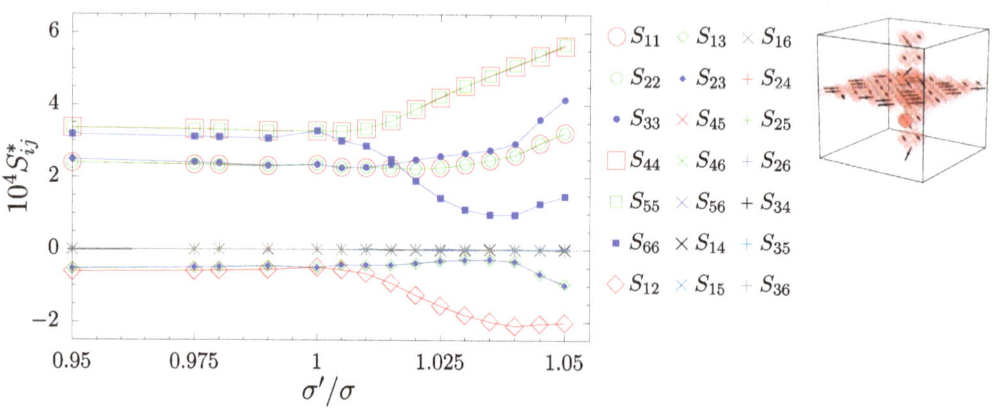

Figure 3. The elements of the elastic compliance matrix S_{ij} for systems with dumbbell inclusions at $p^* = 100$, plotted against σ'/σ.

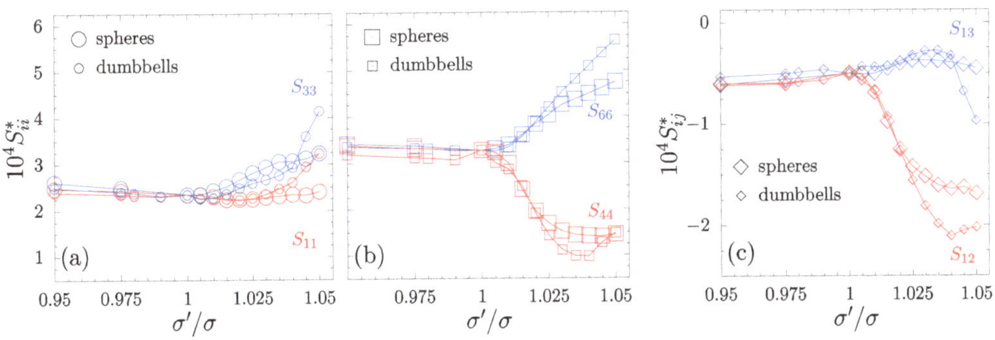

Figure 4. Comparison of elastic compliance matrix elements S_{ij} for systems with inclusions composed of hard spheres [61] (large symbols) and the ones composed of hard dumbbells (small symbols). The colors distinguish elastic compliance matrix elements S_{11} (red) S_{33} (blue) (**a**), S_{44} (red) S_{66} (blue) (**b**), S_{12} (red) S_{13} (blue) (**c**). The respective pairs are equivilent in the cubic system.

The presence of inclusions and the changes in the crystal symmetry are clearly visible in the behavior of the Poisson's ratio of both systems. In Figure 5 PR in typical crystallographic directions has been presented. The notation of the directions in the form ν_{nm}, where n indicates the direction where the load is applied and m is the direction in which the Poisson's ratio is measured. It can be also noted that in the case of cubic symmetry, respective pairs of directions are equivalent. Namely, [100] and [111] are elastically isotropic, so the value of PR in these directions does not depend on m. Furthermore, the PR is invariant on the choice of n as [110] or [101] in the cubic case. It can be seen in Figure 5 that, in the case of inclusions containing a disordered crystalline phase of hard dumbbells, the PR typically increases more at the given $\sigma'/\sigma > 1$. The auxetic properties (an inherent property of an f.c.c. crystal) are removed in both, the [110] and [101] directions at $s = 1.025$. Yet, they are still present in the case of spherical inclusions in the latter direction, and they disappear only at $\sigma'/\sigma = 1.05$. Overall, in the dimer system at $\sigma'/\sigma = 1.025$, the Poisson's ratio in some other directions can be found to be negative, but its value is negligibly small. An interesting difference between the spherical and the disordered molecular inclusions is that the latter significantly decreases Poisson's ratio in the $[110][1\bar{1}0]$ direction when $\sigma'/\sigma > 1.035$ (see also Figure 6). PR reaches -0.235 and shows a significant decrease over -0.06 for the cubic case.

Due to the fact that Poisson's ratio depends on a pair of mutually orthogonal directions, it is hard to demonstrate all the possible changes that occur to the modifications introduced into the crystalline structure. However, it is possible to show the extreme values of PR in the m-directions that correspond to a particular loading direction (n). In the Figure 7 the surfaces of the minimum and maximum PR for all the possible loading directions (parametrized by polar and azimuthal angles θ, φ) are presented for the studied systems at selected σ'/σ. The presented surfaces limit the possible values of PR in the respective directions shown. It is also convenient to see these surfaces plotted in the spherical coordinates. However, as it can be seen on the lower left part of Figure 7, it is necessary to separately plot the surfaces of the maximum (red) and minimum PR. The latter is additionally separated to positive (yellow) and negative (blue) parts. The corresponding scales apply to all the polar plots presented in the figure. For additional details on the presented plots, see [61]. Looking at the plots, one can easily assess the impact of inclusion itself, as well as the impact of the disordered crystalline phase of dumbbells within the inclusion. One can see that the Poisson's ratio reaches its extreme values ($\nu_{\max} = 1$ and $\nu_{\min} = -0.247$) for the system with molecular inclusions, when $\sigma'/\sigma = 1.05$. It can also be seen that hard dumbbells introduced into the inclusion cause a significant increase in the overall auxetic properties of the system, when compared with the pure f.c.c. crystal. This is in contrast with the spherical inclusions, which totally removes the auxetic properties from the system.

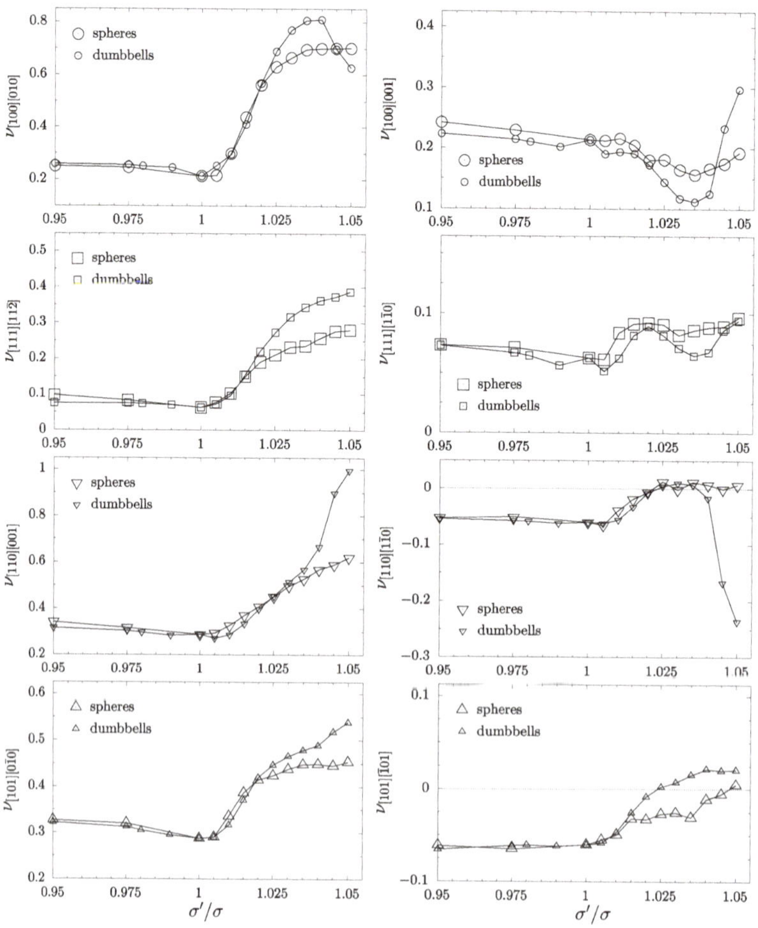

Figure 5. Comparison of the Poisson's ratio ν in the selected, typical crystallographic directions for systems with inclusions composed of hard spheres [61] (large symbols) and the ones composed of hard dumbbells (small symbols).

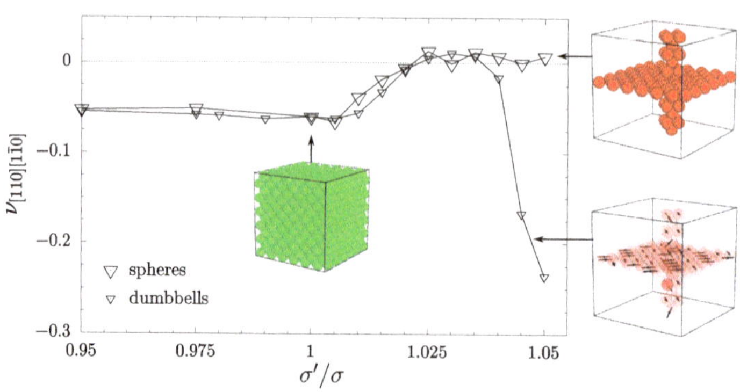

Figure 6. Poisson's ratio in $[110][1\bar{1}0]$-directions for systems with inclusions composed of hard spheres [61] (large symbols) and the ones composed of hard dumbbells (small symbols). The graphical inserts additionally indicate the type of the system.

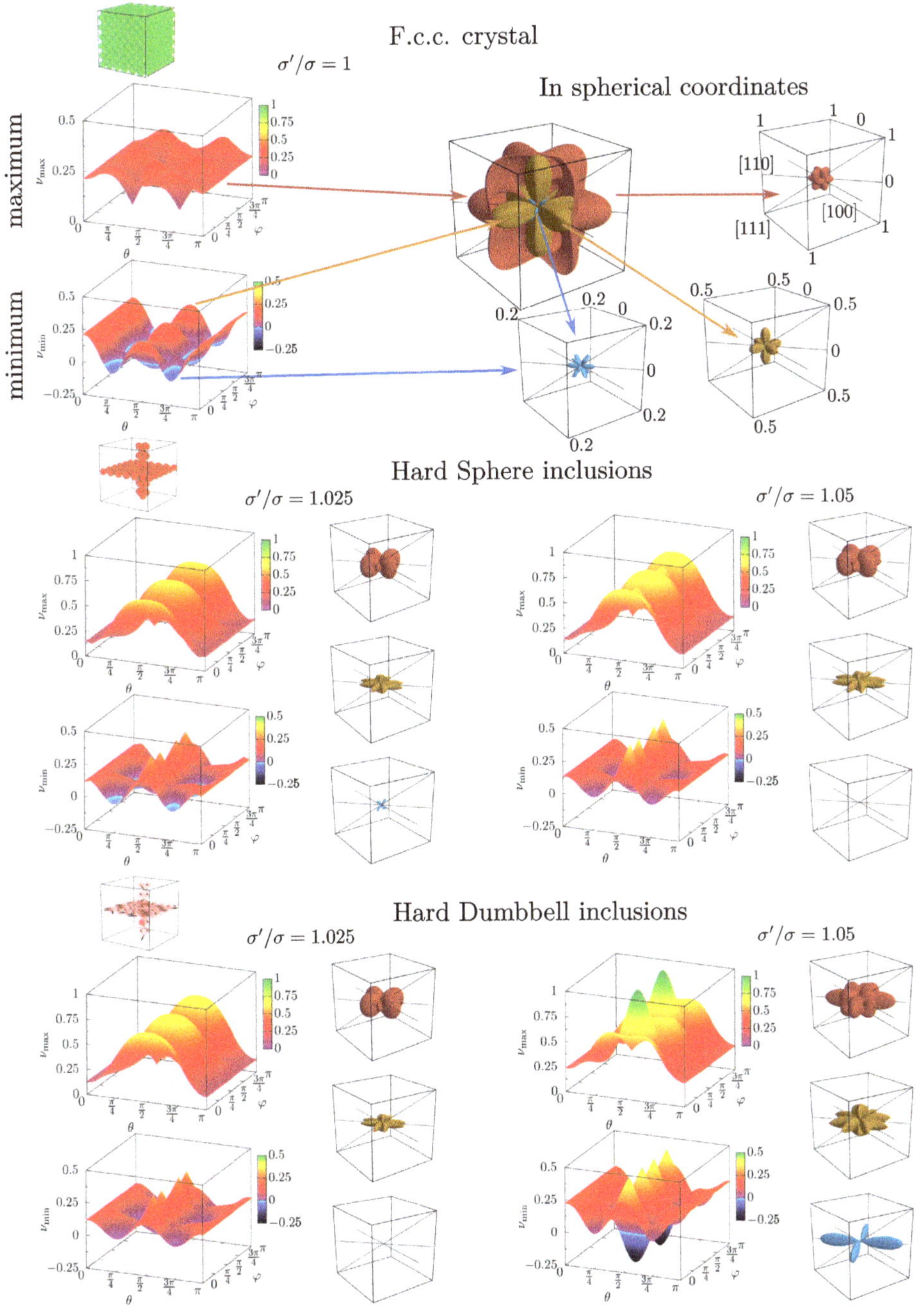

Figure 7. The surfaces of maximum and minimum Poisson's ratio plotted in the Cartesian coordinates as a function of polar and azimuthal angles (θ, φ) that parametrize the loading direction n. The colours on the surfaces indicate the value of the PR defined in the corresponding legend bars. The smaller cubic inserts present the same surfaces plotted in spherical coordinates. Here, the data were divided into maximum PR surface (red) as well as the positive (yellow) and negative (blue) parts of minimum PR surface. The negative part of the maximum PR surface does not exist in this model.

5. Conclusions

The elastic properties of two similar crystal systems of hard particles with structural modifications are compared. The previously studied system of hard spheres containing the combined nanolayer and nanochannel inclusion of hard spheres with different diameters is reviewed and compared with the extended model in which the neighbouring inclusion spheres have been randomly connected in pairs, forming a degenerate crystalline phase of hard dumbbells. It is shown that by introducing the disordered phase into the model system, one can significantly impact its elastic properties. The Monte Carlo simulations in the NpT ensemble showed a significant increase in auxeticity in comparison with the f.c.c. hard sphere crystal. This is also in contrast with the system with nanolayer–nanochannel inclusion composed of unconstrained hard spheres, as in this system, the auxetic properties have been completely removed by the inclusion.

Although the nature of the model interaction used here makes this study purely theoretical, the advances in nanotechnology allow us to fabricate systems even with monoatomic layers [82]. In this context, it is anticipated that ideas presented here will help in designing future real experiments and possibly applications of these effects at a microscale such as sensors or actuators in MEMS systems.

The study presented here focused only on a single size of the inclusion. It is interesting to test the elastic properties of systems with thicker layers and/or wider nanochannels (i.e., different $N_{\rm inc}/N$ ratios) containing disordered phases. The possibility of out-of-plane orientation of molecules may also have an interesting impact on elastic properties. This will be the subject of future works.

Author Contributions: Conceptualization, J.W.N.; Data curation, J.W.N.; Formal analysis, J.W.N.; Funding acquisition, J.W.N.; Investigation, J.W.N.; Methodology, J.W.N.; Project administration, J.W.N., Resources, J.W.N.; Software, J.W.N.; Supervision, J.W.N.; Validation, J.W.N.; Visualization, J.W.N.; Writing—original draft, J.W.N.; Writing—review & editing, J.W.N. All authors have read and agreed to the published version of the manuscript.

Funding: The present work was partially supported by grant No. 2017/27/B/ST3/02955 of the National Science Centre, Poland.

Data Availability Statement: The raw data supporting the conclusions of this article will be made available by the authors upon request.

Acknowledgments: The computations were partially performed at Poznań Supercomputing and Networking Center (PCSS). The author is grateful to Krzysztof W. Wojciechowski from the Institute of Molecular Physics Polish Academy of Sciences for discussions during this work.

Conflicts of Interest: The author declares no conflicts of interest. The funders had no role in the design of the study; in the collection, analyses, or interpretation of data; in the writing of the manuscript, or in the decision to publish the results.

Abbreviations

The following abbreviations are used in this manuscript (listed in order of occurrence in the text):

PR	Poisson's ratio
HS	Hard Sphere
MC	Monte Carlo

References

1. Landau, L.D.; Lifshitz, E.M. *Theory of Elasticity*; Pergamon Press: London, UK, 1986.
2. Baughman, R.H.; Shacklette, J.M.; Zakhidov, A.A.; Stafstrom, S. Negative Poisson's ratios as a common feature of cubic metals. *Nature* **1998**, *392*, 362–365. [CrossRef]
3. Evans, K.E. Auxetic polymers: A new range of materials. *Endeavour* **1991**, *15*, 170–174. [CrossRef]
4. Lim, T.C. *Auxetic Materials and Structures*; Springer: Singapore, 2015. [CrossRef]
5. Lakes, R.S. Foam structures with a negative Poisson's ratio. *Science* **1987**, *235*, 1038–1040. [CrossRef] [PubMed]
6. Wojciechowski, K.W. Constant thermodynamic tension Monte Carlo studies of elastic properties of a two-dimensional system of hard cyclic hexamers. *Mol. Phys.* **1987**, *61*, 1247–1258. [CrossRef]

7. Wojciechowski, K.W. Two-dimensional isotropic model with a negative Poisson ratio. *Phys. Lett. A* **1989**, *137*, 60–64. [CrossRef]
8. Wojciechowski, K.W.; Brańka, A.C. Negative Poisson ratio in a two-dimensional isotropic solid. *Phys. Rev. A* **1989**, *40*, 7222–7225. [CrossRef]
9. Kapnisi, M.; Mansfield, C.; Marijon, C.; Guex, A.G.; Perbellini, F.; Bardi, I.; Humphrey, E.J.; Puetzer, J.L.; Mawad, D.; Koutsogeorgis, D.C.; et al. Auxetic Cardiac Patches with Tunable Mechanical and Conductive Properties toward Treating Myocardial Infarction. *Adv. Funct. Mater.* **2018**, *28*, 1800618. [CrossRef]
10. Kaufman, J.; Farsad, K.; Chinubhai, A.; Bonsignore, C.; Al-Hakim, R. Auxetic Stent Increases Venous Inflow Lumen in an Ovine Animal Model. *J. Vasc. Interv. Radiol.* **2023**, *34*, S46–S47. [CrossRef]
11. Meeusen, L.; Candidori, S.; Micoli, L.L.; Guidi, G.; Stanković, T.; Graziosi, S. Auxetic structures used in kinesiology tapes can improve form-fitting and personalization. *Sci. Rep.* **2022**, *12*, 13509. [CrossRef]
12. Duncan, O.; Foster, L.; Senior, T.; Alderson, A.; Allen, T. Quasi-static characterisation and impact testing of auxetic foam for sports safety applications. *Smart Mater. Struct.* **2016**, *25*, 054014. [CrossRef]
13. Duncan, O.; Shepherd, T.; Moroney, C. Review of auxetic materials for sports applications: Expanding options in comfort and protection. *Appl. Sci.* **2018**, *8*, 941. [CrossRef]
14. Chang, Y.; Hu, H. 3D Fabrics with Negative Poisson's Ratio: A Review. *Appl. Compos. Mater.* **2022**, *29*, 95–108. [CrossRef]
15. Smardzewski, J.; Klos, R.; Fabisiak, B. Design of small auxetic springs for furniture. *Mater. Des.* **2013**, *51*, 723–728. [CrossRef]
16. Kuskun, T.; Kasal, A.; Caglayan, G.; Ceylan, E.; Bulca, M.; Smardzewski, J. Optimization of the Cross-Sectional Geometry of Auxetic Dowels for Furniture Joints. *Materials* **2023**, *16*, 2838. [CrossRef] [PubMed]
17. Prawoto, Y. Seeing auxetic materials from the mechanics point of view: A structural review on the negative Poisson's ratio. *Comput. Mater. Sci.* **2012**, *58*, 140–153. [CrossRef]
18. Saxena, K.K.; Das, R.; Calius, E.P. Three Decades of Auxetics Research—Materials with Negative Poisson's Ratio: A Review. *Adv. Eng. Mater.* **2016**, *18*, 1847–1870. [CrossRef]
19. Lakes, R. Negative-Poisson's-Ratio Materials: Auxetic Solids. *Annu. Rev. Mater. Res.* **2017**, *47*, 63–81. [CrossRef]
20. Ren, X.; Das, R.; Tran, J.P.; Ngo, T. Auxetic metamaterials and structures: A review. *Smart Mater. Struct.* **2018**, *27*, 023001. [CrossRef]
21. Li, X.; Peng, W.; Wu, W.; Xiong, J.; Lu, Y. Auxetic mechanical metamaterials: From soft to stiff. *Int. J. Extrem. Manuf.* **2023**, *5*, 042003. [CrossRef]
22. Alderson, K.; Nazaré, S.; Alderson, A. Large-scale extrusion of auxetic polypropylene fibre. *Phys. Status Solidi B-Basic Solid State Phys.* **2016**, *253*, 1279–1287. [CrossRef]
23. Verma, P.; He, C.B.; Griffin, A.C. Implications for Auxetic Response in Liquid Crystalline Polymers: X-Ray Scattering and Space-Filling Molecular Modeling. *Phys. Status Solidi B-Basic Solid State Phys.* **2020**, *257*, 2000261. [CrossRef]
24. Novak, N.; Vesenjak, M.; Kennedy, G.; Thadhani, N.; Ren, Z. Response of Chiral Auxetic Composite Sandwich Panel to Fragment Simulating Projectile Impact. *Phys. Status Solidi B-Basic Solid State Phys.* **2020**, *257*, 1900099. [CrossRef]
25. Portone, F.; Amorini, M.; Montanari, M.; Pinalli, R.; Pedrini, A.; Verucchi, R.; Brighenti, R.; Dalcanale, E. Molecular Auxetic Polymer of Intrinsic Microporosity via Conformational Switching of a Cavitand Crosslinker. *Adv. Funct. Mater.* **2023**, *33*, 2307605. [CrossRef]
26. Allen, T.; Hewage, T.; Newton-Mann, C.; Wang, W.; Duncan, O.; Alderson, A. Fabrication of Auxetic Foam Sheets for Sports Applications. *Phys. Status Solidi B-Basic Solid State Phys.* **2017**, *254*, 1700596. [CrossRef]
27. Cheng, H.C.; Scarpa, F.; Panzera, T.H.; Farrow, I.; Peng, H.X. Shear Stiffness and Energy Absorption of Auxetic Open Cell Foams as Sandwich Cores. *Phys. Status Solidi B-Basic Solid State Phys.* **2019**, *256*, 1800411. [CrossRef]
28. Duncan, O.; Clegg, F.; Essa, A.; Bell, A.M.T.; Foster, L.; Allen, T.; Alderson, A. Effects of Heat Exposure and Volumetric Compression on Poisson's Ratios, Young's Moduli, and Polymeric Composition During Thermo-Mechanical Conversion of Auxetic Open Cell Polyurethane Foam. *Phys. Status Solidi B-Basic Solid State Phys.* **2019**, *256*, 1800393. [CrossRef]
29. Zulifqar, A.; Hu, H. Development of Bi-Stretch Auxetic Woven Fabrics Based on Re-Entrant Hexagonal Geometry. *Phys. Status Solidi B-Basic Solid State Phys.* **2019**, *256*, 1800172. [CrossRef]
30. Jiang, N.; Hu, H. Auxetic Yarn Made with Circular Braiding Technology. *Phys. Status Solidi B-Basic Solid State Phys.* **2019**, *256*, 1800168. [CrossRef]
31. Zulifqar, A.; Hua, T.; Hu, H. Single- and Double-Layered Bistretch Auxetic Woven Fabrics Made of Nonauxetic Yarns Based on Foldable Geometries. *Phys. Status Solidi B-Basic Solid State Phys.* **2020**, *257*, 1900156. [CrossRef]
32. Tahir, D.; Zhang, M.; Hu, H. Auxetic Materials for Personal protection: A review. *Phys. Status Solidi B-Basic Solid State Phys.* **2022**, *259*, 2200324. [CrossRef]
33. Dudek, K.K.; Martinez, J.A.I.; Ulliac, G.; Kadic, M. Micro-Scale Auxetic Hierarchical Mechanical Metamaterials for Shape Morphing. *Adv. Mater.* **2022**, *34*, 2110115. [CrossRef] [PubMed]
34. Milton, G. Composite materials with Poisson's ratios close to −1. *J. Mech. Phys. Solids* **1992**, *40*, 1105–1137. [CrossRef]
35. Sigmund, O. Tailoring materials with prescribed elastic properties. *Mech. Mater.* **1995**, *20*, 351–368. [CrossRef]
36. Hoover, W.G.; Hoover, C.G. Searching for auxetics with DYNA3D and ParaDyn. *Phys. Status Solidi B-Basic Solid State Phys.* **2005**, *242*, 585–594. [CrossRef]
37. Scarpa, F.; Adhikari, S.; Phani, A.S. Effective elastic mechanical properties of single layer graphene sheets. *Nanotechnology* **2009**, *20*, 065709. [CrossRef]

38. Lim, T.C. Analogies across auxetic models based on deformation mechanism. *Phys. Status Solidi-Rapid Res. Lett.* **2017**, *11*, 1600440. [CrossRef]
39. Shufrin, I.; Pasternak, E.; Dyskin, A.V. Effective properties of layered auxetic hybrids. *Compos. Struct.* **2019**, *209*, 391–400. [CrossRef]
40. Rysaeva, L.K.; Lisovenko, D.S.; Gorodtsov, V.A.; Baimova, J.A. Stability, elastic properties and deformation behavior of graphene-based diamond-like phases. *Comput. Mater. Sci.* **2020**, *172*, 109355. [CrossRef]
41. Grima, J.N.; Evans, K.E. Auxetic behavior from rotating squares. *J. Mater. Sci. Lett.* **2000**, *19*, 1563–1565. [CrossRef]
42. Peliński, K.; Smardzewski, J.; Narojczyk, J. Stiffness of synclastic wood-based auxetic sandwich panels. *Phys. Status Solidi B-Basic Solid State Phys.* **2020**, *257*, 1900749. [CrossRef]
43. Korznikova, E.A.; Zhou, K.; Galiakhmetova, L.K.; Soboleva, E.G.; Kudreyko, A.A.; Dmitriev, S.V. Partial Auxeticity of Laterally Compressed Carbon Nanotube Bundles. *Phys. Status Solidi-Rapid Res. Lett.* **2022**, *16*, 2100189. [CrossRef]
44. Lakes, R.S.; Huey, B.; Goyal, K. Extended Poisson's ratio range in chiral isotropic elastic materials. *Phys. Status Solidi B-Basic Solid State Phys.* **2022**, *259*, 2200336. [CrossRef]
45. Grima-Cornish, J.N.; Attard, D.; Vella-Żarb, L.; Grima, J.N.; Evans, K.E. Boron Arsenate Scaled-Up: An Enhanced Nano-Mimicking Mechanical Metamaterial. *Phys. Status Solidi B-Basic Solid State Phys.* **2022**, *259*, 2200368. [CrossRef]
46. Brizzi, D.D.; Graziosi, S.; Bondin, W.; Grima, J.N.; Airoldi, A. On the behavior of auxetic inserts: A numerical analysis to derive design guidelines. *Smart Mater. Struct.* **2024**, *33*, 075018. [CrossRef]
47. Lim, T.C. Auxetic and Non-Auxetic Metamaterial Model from Interconnected Rotating Parallelograms and Triangles. *Phys. Status Solidi B-Basic Solid State Phys.* **2023**, 2300413. [CrossRef]
48. Lim, T.C. Auxetic System Based on Rotating Hexagons and Triangles. *Phys. Status Solidi B-Basic Solid State Phys.* **2024**, *261*, 2300586. [CrossRef]
49. Strek, T.; Maruszewski, B.; Narojczyk, J.W.; Wojciechowski, K.W. Finite element analysis of auxetic plate deformation. *J. Non-Cryst. Solids* **2008**, *354*, 4475–4480. [CrossRef]
50. Strek, T.; Kedziora, P.; Maruszewski, B.; Pozniak, A.A.; Tretiakov, K.V.; Wojciechowski, K. Finite element analysis of auxetic obstacle deformation and fluid flow in a channel. *J. Non-Cryst. Solids* **2009**, *355*, 1387–1392. [CrossRef]
51. Strek, T.; Matuszewska, A.; Jopek, H. Finite element analysis of the influence of the covering auxetic layer of plate on the contact pressure. *Phys. Status Solidi B-Basic Solid State Phys.* **2017**, *254*, 1700103. [CrossRef]
52. Shepherd, T.; Winwood, K.; Venkatraman, P.; Alderson, A.; Allen, T. Validation of a Finite Element Modeling Process for Auxetic Structures under Impact. *Phys. Status Solidi B-Basic Solid State Phys.* **2020**, *257*, 1900197. [CrossRef]
53. Tretiakov, K.V. Negative Poisson's ratio of two-dimensional hard cyclic tetramers. *J. Non-Cryst. Solids* **2009**, *355*, 1435–1438. [CrossRef]
54. Tretiakov, K.V.; Wojciechowski, K.W. Elastic properties of fcc crystals of polydisperse soft spheres. *Phys. Status Solidi B-Basic Solid State Phys.* **2013**, *250*, 2020–2029. [CrossRef]
55. Tretiakov, K.V.; Wojciechowski, K.W. Auxetic, partially auxetic, and nonauxetic behaviour in 2D crystals of hard cyclic tetramers. *Phys. Status Solidi-Rapid Res. Lett.* **2020**, *14*, 2000198. [CrossRef]
56. Piglowski, P.M.; Narojczyk, J.W.; Poźniak, A.A.; Wojciechowski, K.W.; Tretiakov, K.V. Auxeticity of Yukawa systems with nanolayers in the (111) crystallographic plane. *Materials* **2017**, *10*, 1338. [CrossRef] [PubMed]
57. Hyzorek, K.; Tretiakov, K.V. Thermal conductivity of liquid argon in nanochannels from molecular dynamics simulations. *J. Chem. Phys.* **2016**, *144*, 194507. [CrossRef]
58. Pieprzyk, S.; Yevchenko, T.; Dardas, D.; Brańka, A.C. Phase transitions and physical properties by a color texture analysis: Results for liquid crystals. *J. Mol. Liq.* **2022**, *362*, 119699. [CrossRef]
59. Dardas, D. Survey of Applicable Methods for Determining Viscoelastic Effects in Ferroelectric and Antiferroelectric Chiral Liquid Crystals. *Materials* **2024**, *17*, 3993. [CrossRef]
60. Narojczyk, J.W.; Wojciechowski, K.W.; Tretiakov, K.V.; Smardzewski, J.; Scarpa, F.; Piglowski, P.M.; Kowalik, M.; Imre, A.R.; Bilski, M. Auxetic properties of a f.c.c. crystal of hard spheres with an array of [001]-nanochannels filled by hard spheres of another diameter. *Phys. Status Solidi B-Basic Solid State Phys.* **2019**, *256*, 1800611. [CrossRef]
61. Narojczyk, J.W.; Wojciechowski, K.W.; Smardzewski, J.; Imre, A.R.; Grima, J.N.; Bilski, M. Cancellation of auxetic properties in f.c.c. hard sphere crystals by hybrid layer-channel nanoinclusions filled by hard spheres of another diameter. *Materials* **2021**, *14*, 3008. [CrossRef]
62. Wojciechowski, K.W.; Narojczyk, J.W. Influence of disorder on the Poisson's ratio of static solids in two dimensions. *Rev. Adv. Mater. Sci.* **2006**, *12*, 120–126.
63. Narojczyk, J.W.; Wojciechowski, K.W. Elastic properties of the fcc crystals of soft spheres with size dispersion at zero temperature. *Phys. Status Solidi B-Basic Solid State Phys.* **2008**, *245*, 606–613. [CrossRef]
64. Tretiakov, K.V.; Wojciechowski, K.W. Elasticity of two-dimensional crystals of polydisperse hard disks near close packing: Surprising behavior of the Poisson's ratio. *J. Chem. Phys.* **2012**, *136*, 204506. [CrossRef] [PubMed]
65. Wojciechowski, K.W. Solid phases of two-dimensional hard dumb-bells in the free volume approximation: Crystal-aperiodic-solid phase transition. *Phys. Lett. A* **1987**, *122*, 377–384. [CrossRef]
66. Wojciechowski, K.W.; Frenkel, D.; Brańka, A.C. Nonperiodic solid phase in a two-dimensional hard-dimer system. *Phys. Rev. Lett.* **1991**, *67*, 3168–3171. [CrossRef] [PubMed]

67. Wojciechowski, K.W. Degenerate crystalline phase in a two-dimensional system of hard dimers. *Mod. Phys. Lett. B* **1991**, *5*, 1843–1851. [CrossRef]
68. Tretiakov, K.V.; Wojciechowski, K.W. Poisson's ratio of simple planar 'isotropic' solids in two dimensions. *Phys. Status Solidi B-Basic Solid State Phys.* **2007**, *244*, 1038–1046. [CrossRef]
69. Wojciechowski, K.W.; Brańka, A.C.; Frenkel, D. Monte Carlo simulation of a two-dimensional hard dimer system. *Phys. A* **1993**, *196*, 519–545. [CrossRef]
70. Wojciechowski, K.W. Monte Carlo simulation of highly anisotropic two-dimensional hard dumbbel-shaped molecules: Nonperiodic phase between fluid and dense solid. *Phys. Rev. B* **1992**, *46*, 26–38. [CrossRef]
71. Vega, C.; Paras, E.P.A.; Monson, P.A. Solid–fluid equilibria for hard dumbbells via Monte Carlo simulation. *J. Chem. Phys.* **1992**, *96*, 9060–9072. [CrossRef]
72. Narojczyk, J.W.; Kowalik, M.; Wojciechowski, K.W. Influence of nanochannels on Poisson's ratio of degenerate crystal of hard dimers. *Phys. Status Solidi B-Basic Solid State Phys.* **2016**, *253*, 1324–1330. [CrossRef]
73. Hansen, J.P.; McDonald, I.R. *Theory of Simple Liquids*; Academic Press: Amsterdam, The Netherlands, 2006.
74. Allen, M.P.; Tildesley, D.J. *Computer Simulations of Liquids*; Clarendon Press: Oxford, UK, 1987.
75. Parrinello, M.; Rahman, A. Polymorphic transitions in single crystals: A new molecular dynamics method. *J. Appl. Phys.* **1981**, *52*, 7182–7190. [CrossRef]
76. Parrinello, M.; Rahman, A. Strain fluctuations and elastic constants. *J. Chem. Phys.* **1982**, *76*, 2662–2666. [CrossRef]
77. Wojciechowski, K.W.; Tretiakov, K.V.; Kowalik, M. Elastic properties of dense solid phases of hard cyclic pentamers and heptamers in two dimensions. *Phys. Rev. E* **2003**, *67*, 036121. [CrossRef] [PubMed]
78. Tokmakova, S.P. Stereographic projections of Poisson's ratio in auxetic crystals. *Phys. Status Solidi B-Basic Solid State Phys.* **2005**, *242*, 721–729. [CrossRef]
79. Wojciechowski, K.W. Poisson's ratio of anisotropic systems. *Comp. Meth. Sci. Technol.* **2005**, *11*, 73–79. [CrossRef]
80. Nye, J.F. *Physical Properties of Crystalls, Their Representation by Tensors and Matrices*; Clarendon Press: Oxford, UK, 1957.
81. Narojczyk, J.W.; Tretiakov, K.V.; Smardzewski, J.; Wojciechowski, K.W. Hardening of fcc hard-sphere crystals by introducing nanochannels: Auxetic aspects. *Phys. Rev. E* **2023**, *108*, 045003. [CrossRef]
82. Ali, R.; Saleem, M.R.; Roussey, M.; Turunen, J.; Honkanen, S. Fabrication of buried nanostructures by atomic layer deposition. *Sci. Rep.* **2018**, *8*, 15098. [CrossRef]

materials

MDPI

Review

Survey of Applicable Methods for Determining Viscoelastic Effects in Ferroelectric and Antiferroelectric Chiral Liquid Crystals

Dorota Dardas

Institute of Molecular Physics, Polish Academy of Sciences, 60-179 Poznań, Poland; dorota.dardas@ifmpan.poznan.pl

Abstract: Viscosity, elasticity, and viscoelastic properties are one of the most fundamental properties of liquid crystalline materials; the main problem in determining these properties is the multitude of physical parameters needed to determine the values of elasticity and viscosity constants. In this paper, a number of different measurement methods for the complete characterization of viscoelastic properties for smectic liquid crystalline materials and their mixtures are analyzed, both theoretically and experimentally. The way in which viscoelastic material constants are determined depends mainly on the application/purpose of the materials under study. The subject of this work was to review the methods used to determine viscoelastic effects in ferroelectric and antiferroelectric chiral liquid crystals, their mixtures, composite materials, and even in dielectric systems, which would bear the hallmark of a universal method allowing the application of sufficiently low electric fields. In the case of chiral liquid crystals with ferroelectric and antiferroelectric phases and their subphases, the following assumption applies: fulfilment of Hooke's law (in the case of elastic coefficients) and preservation of laminar flow (in the case of viscosity coefficients).

Keywords: viscoelasticity; ferroelectric liquid crystal; antiferroelectric liquid crystal

1. Introduction

Chiral liquid crystals [1–3] with ferroelectric and antiferroelectric properties have found numerous applications, mainly in optoelectronic devices such as light switches, modulators, displays, and video screens [4–6]. The use of chiral smectics makes it possible to improve greyscale, switching speed, contrast, information content, and other parameters of information imaging devices. Dielectric methods are mainly used in the study of ferroelectric and antiferroelectric properties [7,8], while electro-optical methods have mostly been used to obtain qualitative information [9,10]. It has been noted that the amplitude and phase of the electro-optical response can be additional parameters to help identify sub-phases of liquid crystals, as demonstrated in previous work [11,12]. This extended the possibility of using light modulation depth measurements to verify phase transition temperatures determined by other methods. It was important to show that the intensity of the second harmonic of the electro-optic response is a sensitive, useful indicator of the transition to the paraelectric smectic phase. A method was developed to measure the incident light intensity and the phase shift between the ordinary and extraordinary beam.

Generalized Debye equations have been proposed to describe the relaxation processes recorded by the electro-optical method. It is noted that some components have negative amplitudes [13]. Such a situation has no analogue in dielectric experiments. It was found that the modulation at the fundamental frequency is determined by a linear electro-optic coefficient, while the second harmonic by both coefficients: linear and quadratic. A method was then proposed to make the separation of the first and second harmonics possible. Additionally, a method for determining the electro-optic coefficients was proposed. This provided the possibility to determine the linear a and quadratic c electro-optic coefficients as a function of temperature for ferroelectric and antiferroelectric chiral smectic materials.

An important aspect in the dielectric and electro-optical studies of chiral liquid crystals was the demonstration of their complementarity [13] and the application of linear and nonlinear electro-optical effects [14,15] to the determination of the flexo- and piezoelectric polarization of smectic layers in ferroelectric and antiferroelectric liquid crystals [16]. In the course of scientific activities, due to the application-oriented nature of liquid crystal materials, it has become possible and important to pay particular attention to correct and authoritative methods for the determination of material properties of liquid crystals (not only nematic).

Viscoelastic effects are particularly important and are increasingly and interchangeably referred to as viscoelastic in materials engineering nomenclature [3,17–20].

Four parameters are of the greatest importance in information visualization devices and light modulators from the applications point of view: two elastic coefficients (for the torsional deformations of the smectic director c and the nematic director n) and two rotational viscosity coefficients for these directors [19]. Knowledge of the electro-optic coefficients has extended the research area to include the possibility of determining material parameters such as elasticity or viscosity in application-relevant mesogens. Subsequently, the work on the mechanical properties of chiral liquid crystals was clearly extended in the course of further research, which was directly related to the determination of elasticity and viscosity coefficients in chiral liquid crystals, with a particular emphasis on subphases with ferroelectric and antiferroelectric properties. The study of ferroelectrics was used to test the method and find the relationship between molecular structure and viscoelastic properties, which are important in the design of materials with desired properties and in the composition of mixtures of technical interest. The research of the viscoelastic properties of ferroelectric and antiferroelectric liquid crystals under small deformation conditions is a pioneering study of this kind. The results obtained provide a better understanding of the switching mechanisms in ferroelectric and antiferroelectric liquid crystals. Dielectric and optical methods have been used for deformation detection. Finding a relationship between dielectric and electro-optical measurements extended the capabilities of both methods. Due to the high sensitivity of the electro-optical method, it was possible to apply an extremely low electric field and make measurements in planar and homeotropically oriented samples. Tests on homeotropic samples provided data on the properties of volumetric samples, i.e., data free of surface-induced errors. Rotational viscosity coefficients were determined from dynamic effects (Goldstone and soft modulus relaxation). Detection of relaxation moduli was performed by dielectric and electro-optical methods.

2. Timeline of Measurement Concepts

2.1. Motivation and Aim of the Study

When talking about mechanical properties, one usually thinks of elastic properties in the case of solids and viscous properties in the case of liquids. Liquid crystals combine the properties of both solids and isotropic liquids. This combination of the elastic properties of solids (related to deformation) and the viscous properties of liquids (related to flow) creates some complications in the description of the mechanical properties of liquid crystals, which are therefore called viscoelastic properties. This combination creates a number of interesting phenomena not found in either solids or liquids [21]. In the case of ferroelectric liquid crystals, viscoelastic properties are important from a practical point of view, as they determine the switching rate and threshold voltage in displays. Among the group of liquid crystals whose mechanical properties are well understood, both theoretically and experimentally, we can include nematics [22–24]. Three independent elastic constants and five viscosity coefficients are needed for the mechanical properties of nematic liquid crystals. The above task is much more complicated in the case of smectic liquid crystals [10,25–28]. Here, a much larger number of parameters are needed due to the layered structure (and often lower symmetry). In the most extensive (but still simplified) model of chiral tilted smectics, developed by Dahl and Lagerwall [29], 15 elasticity constants describing only soft, volumetric deformations are included. In these materials, nine independent viscosity coefficients have to be considered, still without taking into account the deformation of the layers [30]. For this reason, the description of the mechanical properties of ferroelectric liquid crystals is very complex, even in its most simplified form. Therefore, some elasticity and viscosity coefficients have not yet been determined experimentally. The only available experimental work concerns the torsional elasticity and rotational viscosity of smectic c- and n-directors. These are coefficients of great technical interest.

2.2. Experimental Determination of Viscosity and Elasticity Coefficients

In isotropic fluids, the viscosity coefficient links the stress to the velocity gradients caused by the flow [31,32]. Similarly, the rotational viscosity coefficient γ links the torque Γ to the angular velocity of the particles:

$$\Gamma = \gamma_{\varphi} \frac{d\varphi}{dt} \quad (1)$$

where φ is the angle of rotation of the molecules.

In the crystalline phase, elasticity coefficients link stresses to deformations. Similarly, in liquid crystals, the elasticity constants are the proportionality coefficients between torque stress and strain. For torsional deformation, we have:

$$\Gamma = -K_{\varphi} \frac{d^2\varphi}{dz^2} \quad (2)$$

The coefficients of rotational viscosity γ and torsional elasticity K_{φ} defined in Equations (1) and (2) are related to the Goldstone mod [27–29], i.e., to changes in the position of the inclination plane of the molecules without affecting the value of the inclination angle. To determine the viscosity coefficient, it is necessary to apply an external force to induce a controlled flow in the fluid [33,34]. Similarly, to determine the elastic coefficient, a controlled deformation should be induced by an electric field. In ferroelectric liquid crystals, an electric field can be used as the source of such an external force. Thanks to its coupling to the spontaneous polarization of smectic layers, flow or deformation can be easily induced [35–37]. In order to develop an experimental method to determine the above coefficients, the two following problems need to be solved:

(1) The value of the angle φ has to be determined from measurements of macroscopic quantities (such as refractive indices or electrical permittivity);
(2) The equation of motion in the c-direction has to be solved, i.e., a well-defined distribution of the azimuthal angle φ on a microscopic scale.

The equation of motion of the c-director has the form [21,38]:

$$-K_{\varphi} \frac{\delta^2\varphi}{\delta z^2} + \gamma_{\varphi} \frac{\delta\varphi}{\delta t} = M \quad (3)$$

where M is the torque acting per unit volume of the liquid crystal. If the external torque is due to an electric field E, then:

$$M = P \times E \quad (4)$$

where P denotes spontaneous polarization.

The solution of the equation of motion (Equation (3)) depends on the shape and amplitude of the applied electric field. The procedure used to determine the γ and K_{φ} coefficients depends on the method by which the deformation of the director c is detected. The methods used to determine the rotational viscosity coefficient γ are summarized in Table 1. Table 2 collects the various methods used to determine the torsional elasticity constant of the c-director.

Table 1. Methods used to determine the rotational viscosity coefficient γ in chiral smectics; τ—relaxation time, τ_s—switching time, A—electrode area, K_φ—torsional elasticity constant for the **c**-director, E—electric field, i—light intensity, P—spontaneous polarizations.

Detection Method	Voltage Shape	Equation	Literature
Light transmission (switching)	Rectangle	$\gamma = \tau_s PE$	[39,40]
Polarization current	Rectangle	$\gamma = 0.556\tau_s PE$	[41]
	Triangle	$\gamma = AP^2 E_{max}/i_{max}$	[42]
Light modulation	Harmonic	$\gamma = Kq^2\tau$	[43,44]
Dielectric constant	Harmonic	$\gamma = Kq^2\tau$	[21,29]

Table 2. Methods for determining the torsional elastic constant K_φ in chiral smectics C*; θ—molecular tilt angle, E_c—critical field strength for the helix unfolding, $q = 2\pi/p$-wave vector of the helix, p—helix pitch, ϑ—inclination of the helix axis, ε_0—electrical permittivity of the vacuum, $\Delta\varepsilon$—dielectric anisotropy, P—spontaneous polarization.

Detection Method	Equation	Literature
Helix deformation Helix unwinding	$K = (16PE_c)/(\pi^2 q^2)$	[45,46]
Light modulation	$K = (PE\theta)/(2vq^2)$	[43,44]
Dielectric constant	$K = P^2/(2\varepsilon_0\Delta\varepsilon q^2\theta^2)$	[47,48]

The number of experimental papers devoted to the measurement of viscosity and elasticity constants in tilted smectic phases is relatively small [40–48]. The methods listed in Tables 1 and 2 have been used to test different materials. However, there are no results obtained for the same material by different methods. In this case, it is practically impossible to compare the methods and their results and their variation can be explained by the material properties.

For the elastic constant K_φ, the experimental results differ by two orders of magnitude [44,46]. The main reason for these discrepancies is probably due to nonlinear effects. The definitions of viscosity and elasticity coefficients assume linear dependence of angular velocity and strain on torque (see Equations (1) and (2)). This is only true for laminar flows and small deformations. Therefore, if large deformations or turbulent flows occur during the measurement, this can cause significant systematic errors. By extension, all methods in which a high electric field is applied, such as the switching method [39–42] or the critical field method [46], have to be used with extreme caution. Low-strain methods [43,44,47,48] seem to be more reliable and can be considered as standard methods.

The study of liquid crystals and their applications has seen significant progress in recent years [49,50]. Traditionally, researchers have focused on applications in liquid crystal displays and optical shutter elements. Nowadays, the spectrum of possibilities includes entirely new and diverse applications in the field of nanotechnology and nanoscience Stabilization of the mesophase is currently being explored. Furthermore, extending and controlling its temperature range through the use of polymers and biopolymers [51] is particularly relevant for phases characterized by a very narrow temperature range, a 0.12 °C range, for example, in so-called "exotic liquid phases". Examples of exotic liquid crystalline phases are the blue phases (BPs), which are characterized by the dispersion of a Kerr constant [52] or in chiral smectic phases (SmC*$_\alpha$, SmC*$_\beta$, and SmC*$_\gamma$) [53]. Functional nanoparticles are also being used to modify the physical properties of liquid crystals by adding ferroelectric [54], antiferroelectric [55], or magnetic particles in a wide variety of shapes and sizes. Hybrid materials, where the liquid crystal is doped with particles that affect optical properties, such as quantum dots, plasmonic, semiconductor, or metamaterials, are also of increasing interest. Self-organization of liquid crystals is used to order and orient matrices, which is currently used in smart glasses or liquid crystal elastomers, among other applications [56].

In most applications of liquid crystal materials, it is important, and sometimes essential, to know a large number of parameters (including elasticity and viscosity coefficients) that influence the behavior of the material under study under controlled and defined conditions [57–60]. The lack of a reference point in relation to chiral liquid crystalline materials, in particular those with ferroelectric and antiferroelectric properties, was my primary motivation to start my research on the determination of the elastic coefficient and rotational viscosity. Knowledge of the material properties of liquid crystals and their influence on the structure, ordering, and characteristics of a given mesophase and neighboring phases is essential for the design of high-performance optoelectronic components such as liquid crystal displays, light modulators, or filters. As mentioned earlier, it is relatively common to determine elasticity and viscosity constants in a strong electric field (which often induces an overshoot of the proportionality range of the elasticity constant and generates turbulent flow, while it should remain laminar); weak electric fields are rarely used, although they are important in assessing the applicability of a material. The type and nature of the phase transformations occurring in the material under investigation can have a decisive influence not only on the value of the determined coefficients but also on their temperature dependence.

The aim of this study was to review the methods used to determine viscoelastic effects in ferroelectric and antiferroelectric chiral liquid crystals, their mixtures, composite materials, and even in dielectric systems that would have the hallmarks of a universal method, allowing the use of sufficiently low electric fields. In the case of chiral liquid crystals with ferroelectric phases, antiferroelectric phases, and their subphases, the following assumption applies: satisfying Hooke's law (in the case of elastic coefficients) and preserving laminar flow (in the case of viscosity coefficients).

2.3. Theoretical Background

To measure any coefficient of elasticity, one must apply a suitable stress to a suitably oriented specimen and then measure the strain caused by that stress. Both stress and strain should be small enough to satisfy the proportionality condition between strain and stress (Hook's law).

For chiral C* smectics, this means that the change in the angle between the c-directions in adjacent smectic layers has to be small compared to the equilibrium value of this angle. Due to the ferroelectric polarizations of the smectic layers in the SmC* phase, the simplest way to introduce deformation is to apply an electric field parallel to the smectic layers. In this case, the low deformation condition means that the electric field strength is much lower than the critical field EC needed for helix unfolding.

Studies of the mechanical properties of the c-director in ferroelectric smectics with an undeformed or weakly deformed helical structure [30] are still scarce. An electro-optical method has been proposed for the determination of the rotational viscosity coefficient γ [39] and the torsional elasticity coefficient B3 [38,61] (according to the notation [29], later papers used the designation K_{φ} and, finally, K) for a smectic c-director, i.e., the projection of a nematic n-director onto the plane of the layer [21]. In the following description, I will consistently use the designation K_{φ} for the elasticity coefficient and γ for the rotational viscosity coefficient. This method satisfies the conditions of low deformation and laminar flow. Furthermore, it was decided to carry out the measurement in homeotropically oriented samples, in which the helix is much less deformed than in the usual planar samples. In the papers [62,63], it was decided to limit ourselves to the determination of the rotational viscosity coefficient γ and the torsional elasticity constant K_{φ}. These coefficients strongly affect the properties of ferroelectric liquid-crystal-based displays. A separate problem is the deformation detection method. Deformation of the molecular distribution causes a change in the macroscopic properties of the sample. As optical properties are relatively easy to detect, the electro-optical method was therefore used to determine deformation. For the numerical estimation of K_φ and γ, it is necessary to assume models linking both the deformation of the molecular distribution to the strength of the electric field and the molecular distribution to the macroscopic optical properties of the sample. In the ideal undisturbed chiral C* smectic layer, the position of the c-director in successive smectic layers is described by:

$$\varphi = \frac{2\pi \cdot z}{p} \tag{5}$$

where φ is the azimuthal angle between the c-direction and an arbitrary x axis of the Cartesian co-ordinate system, z is the co-ordinate in the direction normal to the smectic layers, while p is the helix pitch. When an electric field is applied parallel to the smectic layers, the effective dipole moment of the smectic layers tends to reorient towards the electric field. This reorientation is hindered by elastic and viscous forces. Since the properties of the c-diode are similar to those of the n-diode nematic, we can use the nematodynamic equation to describe the ϕ azimuthal angle [30,61]:

$$K_\varphi \frac{\partial^2 \varphi}{\partial z^2} - \gamma \frac{\partial \varphi}{\partial t} = P_S \cdot E sin\varphi \tag{6}$$

where P_S is the spontaneous polarization and $E = E_0 \cos t$ denotes the electric field strength with an angular frequency of ω applied to the sample. For small deformations, the steady-state solution of Equation (5) is of the form:

$$\varphi = qz + \varphi_0 \cdot \cos(\omega t + \beta) \cdot \sin qz \tag{7}$$

where $q = 2$ and the amplitude of the change in angle φ is φ_0:

$$\varphi_0 = \frac{P_S \cdot E_0}{\gamma} \cdot \frac{1}{\sqrt{\frac{1}{\tau^2} + \omega^2}} \tag{8}$$

and $\beta = arctan\omega t$, where τ is the relaxation time:

$$\tau = \frac{\gamma}{K_\varphi \cdot q^2} \tag{9}$$

The second term of Equation (7) describes the response of the chiral smectic C* to the applied electric field AC. The angle changes are also periodic but shifted by an angle β. The deformation of the helix described by Equation (8) causes changes in the dielectric and optical properties of the chiral C* smectic. One easily detectable change is the shift in the optical axis. When the inclination angle of the φ_0 molecule is small, the relationship between the amplitude of the azimuthal angle and the amplitude of the inclination of the optical axis has the following approximate form [61]:

$$\vartheta_0 = \frac{1}{2} \varphi_0 \cdot \theta \tag{10}$$

Thus, measuring the amplitude of the slope of the optical axis at different frequencies makes it possible to determine $K_{\vartheta 0}$ and γ using Equations (8) and (9). When γ << 1/τ, then ϑ0 does not depend on frequency and, from Equations (8)–(10), we obtain [62]:

$$K_\varphi = \frac{P_S \cdot E_0 \cdot \theta}{2\vartheta_0 \cdot q^2} \tag{11}$$

From the frequency dependence of the ϑ0, the relaxation time of the τ can be calculated, followed by the viscosity of the γ [63]:

$$\gamma = K_{\varphi} \cdot q^2 \cdot \tau = \frac{P_S \cdot E_0 \cdot \theta}{2\vartheta_0} \tau \tag{12}$$

where τ—relaxation time, τ_s—switching time, A—electrode area, K_{φ}—torsional elasticity constant for the c-director, E—electric field, i—light intensity, P , P_S—spontaneous polarisation, $q = \frac{2\pi}{p}$—wave factor of helix, p—pitch, and ϑ_0—azimuth angle.

As shown earlier, the field-induced collective motion of molecules in smectic ferroelectric liquid crystals with helical superstructure is usually studied by analyzing the fluctuation of the azimuthal angle ϑ along the z axis normal to the smectic layers. In addition, it is assumed that the molecules do not interpenetrate each other and the smectic layers are well defined. Then, the rotational motion of the molecules induced by an external electric field varying sinusoidally with amplitude E_0 and angular frequency ω is described by Equation (6) [1,9].

In the case of antiferroelectric liquid crystals, the polarization of each pair of adjacent smectic layers is almost abolished but, due to the presence of a helical structure, there is still a nonzero value [1]. Since the thickness of the smectic layer l is much smaller than the helix pitch p, the absolute value of the remaining local polarization associated with a pair of smectic layers can be determined by the relation:

$$\delta P_S = 2\pi \frac{l}{p} P_S \tag{13}$$

Given the solution of the equation of motion of a pair of molecules belonging to adjacent smectic layers, it can easily be shown, by analogy with Equation (11), that, for antiferroelectric liquid crystals, the solution can be obtained simply by replacing P_S in Equation (6) by δP_S [64]. Then, using Equations (10), (11) and (13) and considering that $l = l_0 cos\theta$, where l_0 is the length of the molecule (the linear size of the molecule measured in the direction parallel to the n-director), we obtain:

$$K_a = \frac{P_S \cdot E_0 \cdot l_0 \cdot sin2\theta}{4\vartheta_0 \cdot q^2} \tag{14}$$

The right side of the above relationship includes parameters that can be determined experimentally. To calculate ϑ_0, the method of measuring the linear depth of light [29] was used, which will be presented in more detail when describing the experiment. Thus, this equation allows us to calculate the torsional elasticity constant for an antiferroelectric liquid crystal in a weak electric field [65,66].

The change in the orientation of the optical axis $\Delta\alpha$ in a weak electric field can be expressed as [2]:

$$\Delta\alpha = aE_0 \tag{15}$$

with a coefficient a independent of E_0. The linear electro-optical coefficient a [12] describing the change in orientation of the optical axis under the influence of a weak electric field E that does not destroy the helical structure can be expressed as:

$$a = \frac{1}{2} \frac{P_S \sin\theta}{Kq^2\sqrt{1+\omega^2\tau^2}} \tag{16}$$

where $q = 2\pi/p$, p is the helix pitch, θ the angle of inclination of the molecules in the smectic layer (in the ferroelectric smectic phase C*, chiral molecules are spontaneously inclined with respect to the axis normal to the layer), and τ denotes the relaxation time of the Goldstone modulus (9). Solving Equation (6) for small deformations, after taking into account Equation (16), for the condition $\tau << 1$, the expression for the interlayer elasticity factor K_φ [67–69] can be obtained in the form:

$$K_\varphi = \frac{1}{8\pi^2}\frac{P_S \cdot p^2 \cdot sin\theta}{a} \tag{17}$$

Equation (17) allows calculation of the coefficient K_φ when the following parameters are known: a (linear electro-optical coefficient), P_0 (local spontaneous polarisation), p (pitch), and θ (angle of inclination of molecules). In addition, from Equations (9) and (17), if τ is known, the rotational viscosity can be determined [70]:

$$\gamma = \frac{P_S \cdot \theta}{2a}\tau \tag{18}$$

where τ is the relaxation frequency of the Goldstone modulus, that is, the relaxation mod associated with fluctuations of the long axis of the molecules in the azimuthal direction with respect to the normal to the smectic layer. The formulas in Equations (17) and (18) are the basis for determining the viscoelasticity in ferroelectric helical liquid crystals in the small deformation limit and the laminar flow limit. In the case of antiferroelectric liquid crystals, after considering Equation (13), we obtain [71]:

$$K_a = \frac{1}{8\pi^2}\frac{\delta P_S \cdot p^2 \cdot sin\theta}{a} \tag{19}$$

and:

$$\gamma_a = \frac{\delta P_S \cdot \theta}{2a}\tau \tag{20}$$

The formulas in Equations (19) and (20) are the basis for determining the viscoelasticity in ferroelectric liquid crystals with helical structure in the limit of small deformations and the limit of laminar flow.

An expanded list of available methods for determining the viscoelasticity of K_φ and K_a in chiral smectics with ferroelectric and antiferroelectric properties is presented in Table 3, while an expanded list of available methods for determining the rotational viscosity of γ and γ_a is presented in Table 4. K_φ and K_a denote the coefficients of elasticity in the ferroelectric and antiferroelectric phases, while γ and γ a denote the rotational viscosity in the ferroelectric and antiferroelectric phases, respectively.

Table 3. Methods used to determine the elastic coefficient K_φ and K_a in chiral smectics with ferroelectric and antiferroelectric properties. The gray color indicates the new methods described in [65,66,69,71].

Detection Method	Equation	Literature
Helix deformation Helix unwinding	$K_\varphi = (16PE_c)/(\pi^2 q^2)$	[45,46]
Light modulation	$K_\varphi = (PE\theta)/(2vq^2)$	[43,44]
Dielectric constant	$K_\varphi = P^2/(2\varepsilon_0 \Delta\varepsilon q^2 \theta^2)$	[47,48]
Light modulation	$K_\varphi = \frac{P_S \cdot E_0 \cdot \theta}{2\theta_0 \cdot q^2}$	[65]
Light modulation	$K_a = \frac{P_S \cdot E_0 \cdot l_0 \cdot sin2\theta}{4\theta_0 \cdot q^2}$	[66]
Light modulation	$K_\varphi = \frac{1}{8\pi^2}\frac{P_S \cdot p^2 \cdot sin\theta}{a}$	[69]
Light modulation	$K_a = \frac{1}{8\pi^2}\frac{\delta P_S \cdot p^2 \cdot sin\theta}{a}$	[71]

Table 4. Methods used to determine the viscosity coefficient of γ and γ_a in chiral smectics with ferroelectric and antiferroelectric properties. Gray color indicates new methods described in [62,70,71].

Detection Method	Voltage Shape	Equation	Literature
Light transmission (switching)	Rectangle	$\gamma = \tau_s PE$	[39,40]
Polarisation current	Rectangle	$\gamma = 0.556\tau_s PE$	[41]
	Triangle	$\gamma = AP^2 E_{max}/i_{max}$	[42]
Light modulation	Harmonic	$\gamma = K_\varphi q^2 \tau$	[43,44]
Dielectric constant	Harmonic	$\gamma = K_\varphi q^2 \tau$	[21,29]
Light modulation	Harmonic	$\gamma = \frac{P_S \cdot E_0 \cdot \theta}{2\vartheta_0}\tau$	[63]
Light modulation	Harmonic	$\gamma = \frac{P_S \cdot \theta}{2a}\tau$	[70]
Light modulation	Harmonic	$\gamma_a = \frac{\delta P_S \cdot \theta}{2a}\tau$	[71]

3. Materials and Methods

The subjects of this study were smectic chiral liquid crystals, differing in the sequence of phase transitions. The first of these was 4′-methylbutylxy phenyl-4-acetylxy-benzoate material, designated as C8 [62,63,68]—showing only two mesophases. The ferroelectric phase appears in it immediately after softening; then, we observe a transition to the paraelectric phase of the twisted nematic. This material has been tested by various methods used to date [68]. Another commercially available selected material was 4-(n-hexylxy phenyl)-1-(2-fuethyl butyl) biphenyl-4-carboxylate, called Ce-3 [69,70], whose phase sequence was analogous to C8, and 4-(2-methylbutyl) phenyl-4-n-octylbiphenyl-4-carboxylate, Ce-8 [69,70], a ferroelectric liquid crystal, which, in turn, exhibits a very rich polymorphism; from the smectic phase C* during heating, we observe a phase transition to the non-spiral, paraelectric smectic phase A*, then to the chiral, nematic phase N*, and then to the blue phase, whose periodicity is formed by a superstructure of three-dimensional defects. In the group of materials that exhibit antiferroelectric properties is 4-(1-methyl-heptyloxycarbonyl) phenyl 4′-(3-butanoyloxy propyl-1-oxy) biphenyl-4-carboxylate, abbreviated as D12 [65,66], synthesized at the Military University of Technology in Warsaw, Poland (see Table 5). This material is characterized by a phase transition directly from the antiferroelectric phase SmC* to the paraelectric phase A*. The last material presented is 4-(1-methylheptyloxycarbonyl)phenyl 4-acetylxybiphenyl-4-carboxylate, a commercially available material (Military University of Technology, Warsaw, Poland, Sigma-Aldrich, Saint Louis, MO, USA) exhibiting rich polymesomorphism, with the acronym MHPOBC [71]. Phase transitions from the antiferroelectric SmC*$_A$ and ferroelectric SmC* phases to the paraelectric SmA* phase take place through so-called exotic phases with ferroelectric properties, respectively, SmC*$_\gamma$ and SmC*$_\alpha$.

Table 5. Liquid crystalline materials studied.

Short Name	Chemical Structure	Phase Sequence
C8	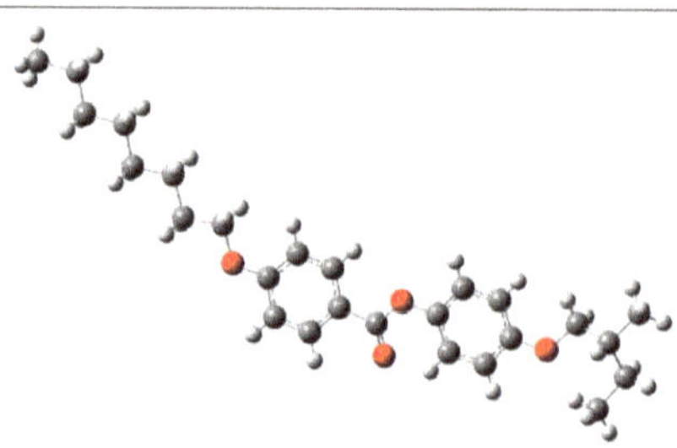	Cr 48 (SmIV 12 SmIII 21) SmC* 43.1 SmA* 59 Iso

Table 5. *Cont.*

Short Name	Chemical Structure	Phase Sequence
Ce-3	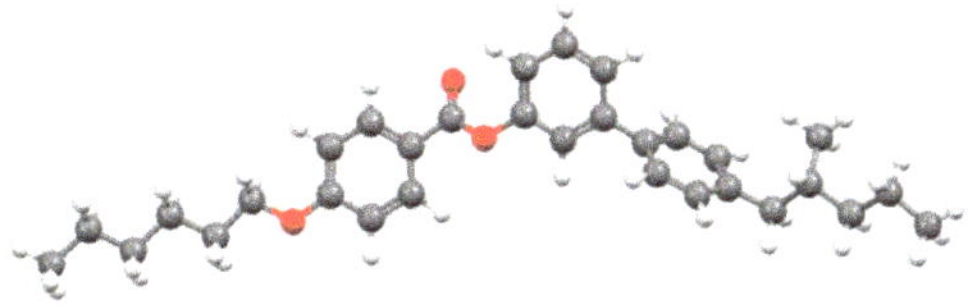	Cr 65 SmC* 77.5 N* 162 Iso
Ce-8	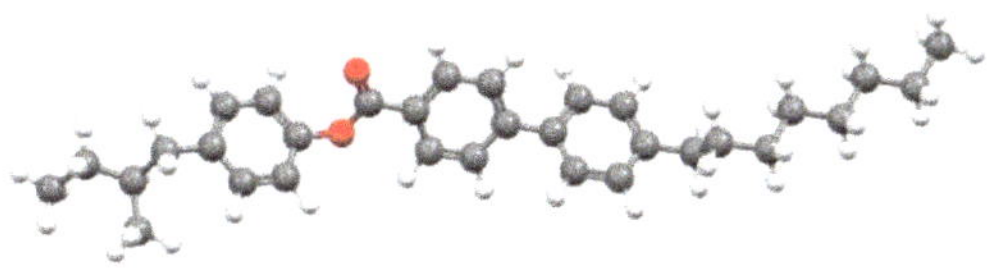	Cr 48 SmG 63.3 SmJ 64.7 SmF 66.7 SmI 69 SmC* 85 SmA* 135.4 N* 140.7 BP 141 Iso
Ce-3/8 mixture	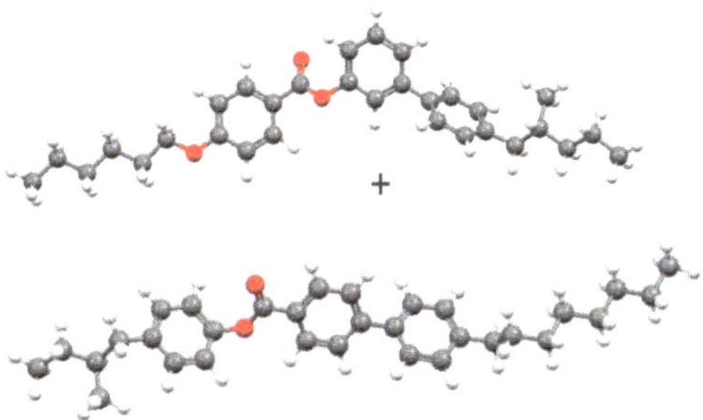	Cr 39.6 SmG 56 SmJ 65 SmF 67 SmC* 86 SmA*124 N*145.5 BP 147 Iso
D12	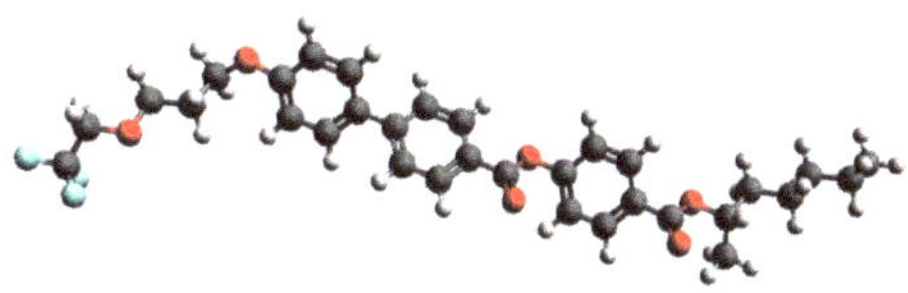	Cr 111.5 SmC_A* 125 SmA* 136 Iso
MHPOBC	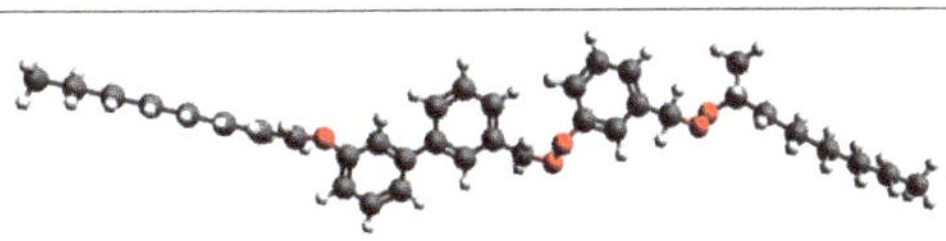	Cr 59.0 SmC^*_A 119.5 SmC^*_γ 120.5 SmC^*_β 121.9 SmC^*_α 123.0 SmA* 150.2 Iso

A typical electro-optical measurement cell consists of two glass plates separated by spacers. This ensures that the inserted layer of liquid crystal has a well-defined thickness. For measurements with planar ordering, commercial cells manufactured by Instec (Instec Inc., Boulder, CO, USA) with a thickness of 5 μm, AWAT (Warsaw, Poland) with a thickness of 5 μm, and EHC (EHC Co., Ltd, Tokyo, Japan) with a thickness of 12 μm, 30 μm, and 110 μm were used. The conductive surfaces (electrodes) are made of indium tin oxide and coated with a polymer orientation layer. Thin electrical wires are attached to the electrodes. The cells used for measurements with homeotropic ordering were prepared using two strips of aluminum foil glued to a glass plate at a distance of about 1–1.6 mm. Another glass plate was attached to the top of the electrodes. Both glass plates were coated with a layer of surfactant (hexadecyl-trimethylammonium bromide) to achieve homeotropic boundary conditions. The temperature suitable for filling the cell is that of an isotropic liquid. A drop of liquid is placed near the gap between the slides, which, due to the capillary effect, fills the entire cell. We can use the cell prepared in this way for thermo-optical, electro-optical, dielectric, and phase transition observations. In order to obtain a homogeneous orientation of the smectic layers of the liquid crystal, perpendicular or parallel to the electrodes, various procedures are often used. Examples include very slow cooling from the isotropic phase, introducing a temperature gradient near the phase transition, using a suitably modulated voltage, or applying a DC voltage. There is no single universal method. Thin cells provided better ordering of smectic layers but strongly deformed the helix. On the other hand, thick samples did not significantly affect the structure of the helix but caused worse ordering of the layers. The ordering of the sample was checked by using a polarizing microscope.

In order to determine the rotational viscosity coefficient γ and the torsional elasticity constant K_φ for chiral smectics with ferroelectric and antiferroelectric properties, the following quantities have to be measured:

a. The spontaneous polarization of P_S;
b. The pitch p;
c. Inclination angle θ;
d. Linear electro-optical coefficient a;
e. The relaxation time τ.

Analyzing the equations presented, it is not difficult to see that the value of elasticity and viscosity coefficients in liquid crystals with ferroelectric and antiferroelectric properties is influenced by spontaneous polarization, the tilt angle of the molecules, and, in particular, the helix pitch. Spontaneous polarization identifies the coupling of the applied electric field to the dipole moment of the molecules, the layers, and the volume response and the tilt angle of the molecules influences the electro-optical response of the system under study. However, it is worth emphasizing that it is the precise, conformal measurement of the helix pitch that can determine the validity of the results obtained. This is mainly due to the quadratic dependence of the wave vector of the helix for both the elastic coefficient and the rotational viscosity. For this reason, the determination of the helix pitch in smectics with properties should be carried out under volumetric conditions (bulk). In the case of chiral smectics, such conditions are met when the planar arrangement allows an undeformed helix to unfold, which requires a large cell thickness (and an applied electric field) appropriate to the pitch of the helix or a homeotropic arrangement that allows the helix to unfold unconstrained, parallel to the limiting surfaces.

a. Spontaneous polarization was measured by a standard method, using a Diamant-Pepinsky bridge [67]. Spontaneous polarization [62,63,68] was determined in homeotropic cells, while, for [66,69–71], in planar cells.
b. The pitch was measured in homeotropically oriented samples using the following methods: spectroscopic [62,68–70] and the Cano method in a wedge sample [62,68] and in a homeotropically oriented free liquid crystal droplet [71]. The aforementioned methods give comparable results under the assumption that the refractive index of the common ray is 1.4.
c. The tilt angle of the molecules was measured using the alternating field method. An electric field of rectangular shape with a fairly high frequency (from tens to hundreds of hertz) was used here. Light intensity was measured using a photodiode and observed on an oscilloscope.
d. Changing the position of the optical axis causes a change in the intensity of light passing through the sample. The intensity of the light is recorded by a photodiode. At the output of the preamplifier, a signal proportional to the light intensity is obtained. The signal modulated in this way contains, in addition to a constant component, a variable component, called the electro-optical response. In addition, changes in the angle α can be simulated by rotating the microscope table along with the sample by a small angle αK [65,66]. In the work [69–71], an additional modification of this method was used. In the experiment, the constant component of the intensity of light passing through the sample was measured simultaneously with the variable component. During the measurement of the electro-optical response, the sample made oscillations with an amplitude of 0, 14°. The frequency of the oscillations was determined by the speed of rotation of the synchronous motor putting the system into motion and was 6.25 Hz. Calibration of the electro-optical response using the

above-described system made it possible to obtain values of the light modulation depth in absolute units.

e. Relaxation time was determined from measurements of the real and imaginary components of the electro-optical response as a function of frequency.

The main principles of measuring viscoelastic effects using the electro-optical method and step-by-step linear measurement conditions are as follows: first, the material to be tested by the capillary method is placed in an electro-optical cell with a defined planar order and thickness. The sample prepared in this way is placed in a heating oven connected to a measuring system with the possibility of precise temperature control and dielectric, electro-optical measurements, and electro-optical response detection with the simultaneous possibility of observing the material with a polarizing microscope. This should be followed by a procedure for homogeneous ordering of the sample; the most effective method has proven to be very slow cooling from the isotropic phase in a slowly varying electric field. When the material is in a smectic, ferroelectric phase, we position it at 22.5 deg between crossed polarizers, where the electro-optical response shows a maximum. To determine the linear electro-optical coefficient a, the real and imaginary components of the electro-optical response should be measured as a function of temperature; to determine the relaxation time, it should also be measured as a function of frequency [12,66]. Both relationships are measured simultaneously using a lock-in amplifier during the oscillation of the microscope table caused by the stepping motor with the photodiode, allowing the electro-optical response to be calibrated and absolute values to be obtained. Then, without taking the sample out of the oven, with the same temperature controller, the spontaneous polarization P_S in the Diamant system and the tilt angle θ of the molecules are measured using an oscilloscope. A separate experiment is required to determine the pitch in the volume cell (with homeotropic ordering, preferably in a free droplet). The measurement requires the use of a spectrophotometer. The results obtained in this way are used to determine the coefficient of elasticity and rotational viscosity.

4. Results and Discussion

On the basis of measurements made by complementary methods of many physical quantities, it was possible to determine the elastic coefficient K_φ and rotational viscosity γ in liquid crystalline materials with ferroelectric properties [62,63]. It was assumed that the parameters corresponding mostly to the behavior of volume samples should have at least deformed helical structure. Disturbances of the helix of chiral smectics can be caused by too strong anchoring conditions in thin planar samples, as well as by the use of too strong electric fields that cause a complete pitch disruption and saturation state in thicker samples. The results obtained for the ferroelectric C8 material obtained under conditions similar to the bulk sample (homeotropic ordering) and in weak electric fields (guaranteeing the linearity and reversibility of the studied effects, according to Hooke's law, and with a relatively low flow velocity of the liquid crystal guaranteeing its laminarity [62,63]) are presented in Figures 1 and 2.

The rotational viscosity coefficient γ disappears at the SmC*–SmA* phase transition. This is not surprising, since this parameter describes the properties of the c-director, which disappears at the transition to a non-tilted phase that no longer has a helicoidal structure. Comparison of our experimental results with literature data is difficult, since data on material constants are very sparse. The scatter of measurement results is very large (more than two orders of magnitude). The literature data are mostly obtained for flat samples, and we suppose that the results are strongly dependent on the geometry of the sample. In addition, it is difficult to compare results for different materials. The values of the rotational viscosity coefficient obtained in this article agree fairly well with some literature data [21,33,72] but are clearly very different from many others obtained in planar orientation; see, for example, [73–75].

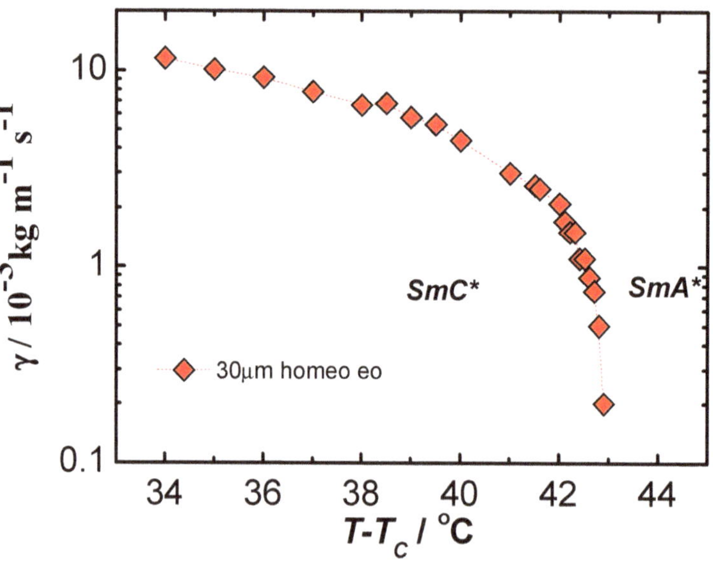

Figure 1. Temperature dependence of the rotational viscosity coefficient γ in a ferroelectric liquid crystal C8.

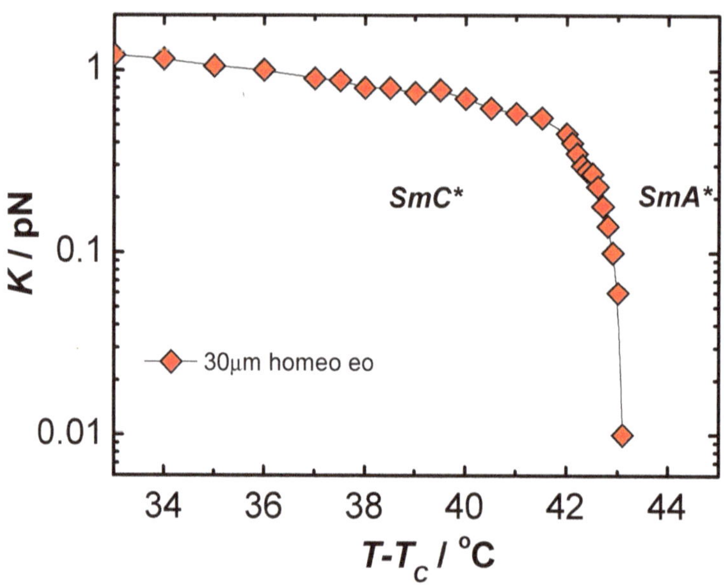

Figure 2. Temperature dependence of the elastic coefficient K_{φ} in a ferroelectric liquid crystal C8.

When analyzing the results, special attention should be paid to determining the pitch p to ensure correct values of the elastic constant K_{φ}. The pitch should be measured with high accuracy, since its square appears in Equation (7). The stresses present in the helix can affect the response of the optical axis of the sample due to the electric field [76]. The best way to avoid the influence of stresses on the determined values of the elastic constant is to measure the inclination of θ and the inclination of the optical axis ϑ_0 at the same time or at least in the same sample geometry. Therefore, the results of measurements of the inclination of molecules in homeotropic samples were used to calculate K_φ. The elasticity constant K_φ vanishes at the SmC*–SmA* phase transition. Comparison of our experimental results with literature data is difficult because, so far, measurements of the volume elasticity coefficient K_φ in weak fields have been described only in two papers [76,77].

The K_φ values obtained in these works are similar to those presented here (0.5×10^{-12} N, 1×10^{-12} N, and $0.9 \times 10{-}12$ N, respectively, at temperatures 3×10^{-12} N below T_{CA}). However, data obtained for planar samples are much more scattered (for example, [72,73,75]). Most of these are based on dielectric measurements, which are strongly dependent on the sample geometry [26,44,61].

However, it is difficult to make any conclusive statements because, here too, the results apply to different materials. In summary, values of viscoelastic coefficients were obtained using a weak electric field. All measurements were made in cells with homeotropic ordering, which allows one to get significantly closer to the volume properties of the material under study. The results obtained look promising, so it was next decided to test the material 4′-methylbutylxy phenyl-4-octyloxy-benzoate, called C8, a liquid crystal with ferroelectric properties by available methods, and thus compare the results obtained for one material versus temperature. The results of such an analysis are shown in Figure 3 for the value of the obtained elastic coefficient K_φ and in Figure 4 for the rotational viscosity γ.

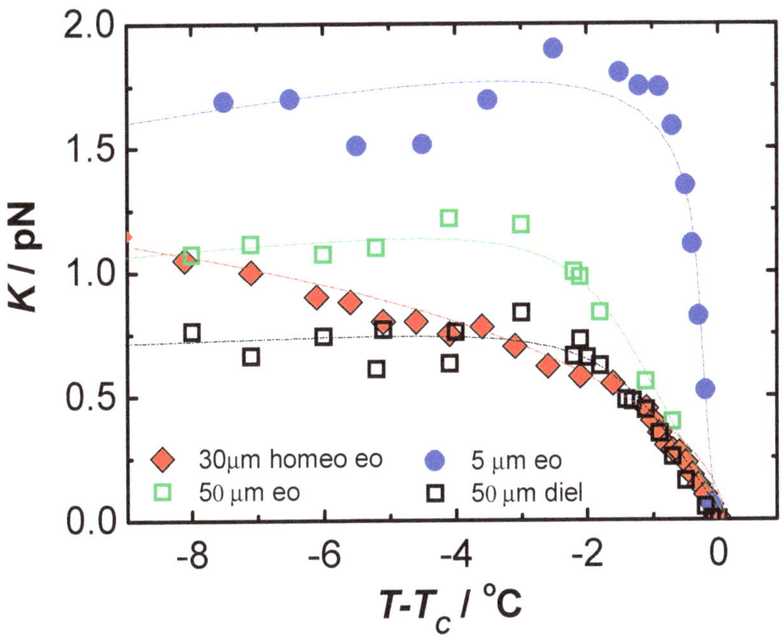

Figure 3. Temperature dependence of the elastic coefficient K_φ in ferroelectric liquid crystal C8 obtained by various methods. In the diagram, "eo" denotes electro-optical methods and "diel" dielectric methods. Thicknesses of planar samples, 5 µm and 50 µm; thickness of homeotropic sample, 30 µm.

Figure 3 clearly shows that the experimental conditions (primarily thickness and type of ordering) have a strong influence on the measurement results [68]. The results differ not only in values but also in the shape of the temperature dependence. It is worth noting that the temperature dependencies are rather irregular for planar-ordered samples. A similar behavior was observed for the rotational viscosity coefficient γ, shown in Figure 4. It was supposed that surface interactions are the main reason for the differences in the results obtained under different experimental conditions. The largest values of K_φ coefficients were obtained for a thin (5 μm), planar-ordered sample, where surface interactions play the most important role.

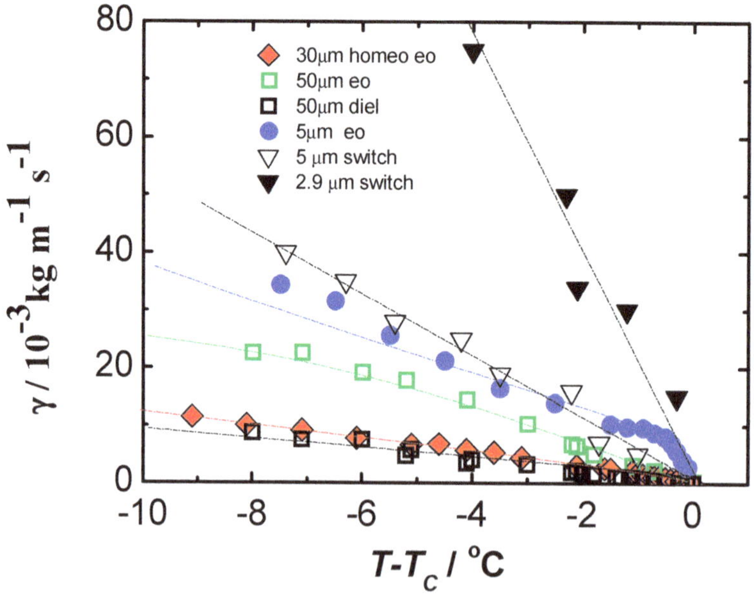

Figure 4. Temperature dependence of the rotational viscosity coefficient γ in ferroelectric liquid crystal C8 obtained by various methods. In the diagram, "eo" denotes electro-optic methods, "diel" dielectric, and "switch" switching methods. Thicknesses of the planar samples, 2.9 μm, 5 μm, and 50 μm; thickness of the homeotropic sample, 30 μm.

Since the helical structure of the smectic SmC* phase is not always compatible with planar ordering, the properties of the cell greatly affect the measurement results. In the case of homeotropic orientation, the helical structure and other material properties have less influence on the measurement results and are less surface-dependent. The smectic layers of the homeotropically oriented sample are parallel to the cell walls, and the helical structure is only slightly deformed near the surface. We believe that the results obtained in thick homeotropic cells better reflect the bulk properties of the chiral smectic phase C*. Therefore, we assume that we can consider these results as reference results. As can be seen in Figure 3, there is some correlation between the dielectric results obtained for the 50 μm planar sample and the 30 μm homeotropic sample. The discrepancy may be due to the different temperature dependence in the planar and homeotropic sample [77]. The helix pitch should always be measured with high accuracy [76], since its square appears in theoretical formulas for viscoelasticity coefficients [62,64]. It is obvious that the stresses deforming the helix can affect the response of the optical axis to the applied electric field and consequently cause fluctuations in the value of the K_{φ} parameter.

Figure 4 shows the results of rotational viscosity γ determined by various methods. Abbreviations were used, where "diel" stands for dielectric methods, "eo" electro-optical, "switch" switching method, and "homeo eo" our proposed method for testing a near-volumetric sample. As you can see, the discrepancy in results is significant. The switch method used to determine the rotational viscosity coefficient γ requires special attention. It is the most widely used method. As Figure 4 shows, the results obtained by this method differ significantly from the results of all other methods. Moreover, the dependence on cell thickness is also the greatest. Therefore, the results obtained by the switching method should be treated with great caution.

Publication [68] focused on methods for determining viscoelastic properties under conditions of low influence of surface interactions and in the presence of small deformations and compared them with switching experiments induced by turbulent flow. Another factor that strongly influences the experimental results is the anchoring of the surface. As a rule, elasticity and viscosity coefficients were measured in thin, planar, surface-stabilized samples. Surface interactions are quite weak in homeotropic cells. Using a typical ferroelectric liquid crystal, it has been shown that the correct values of the aforementioned material constants in the chiral smectic C* phase can be best determined using thick and homeotropically ordered samples. However, this method has a considerable limitation. Namely, obtaining homogeneous ordering in a thick, homeotropic sample without using a very strong electric field is extremely difficult. Determination of material constants in volumetric samples is important for material characterization, while the most common applications of ferroelectric and antiferroelectric liquid crystals are in displays, modulators, or smart films, where the material is surface-stabilized.

Figure 5 collects the results obtained for various liquid crystal materials with ferroelectric properties obtained by the new electro-optical method described in [65] using weak electric fields.

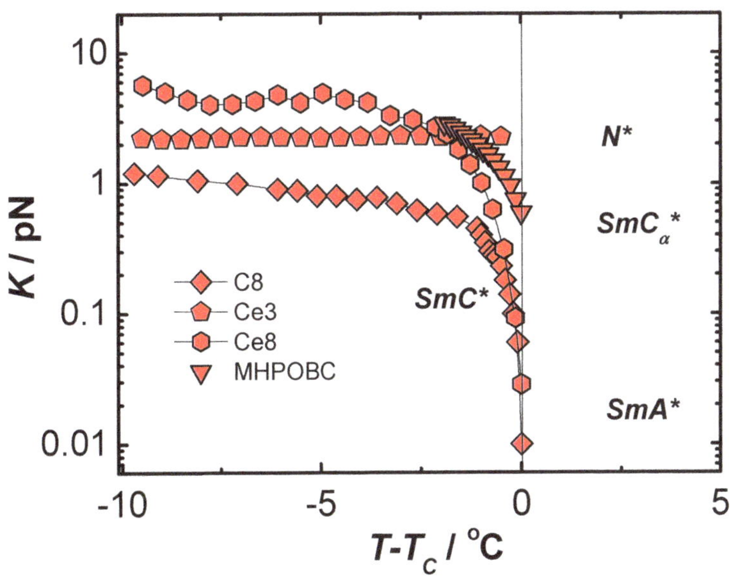

Figure 5. Temperature dependence of the elastic coefficient K_φ in chiral liquid crystals with a ferroelectric phase: C8 (rhombs), Ce3 (diamonds), Ce8 (hexagons), and MHPOBC (triangles) obtained by electro-optical method using low electric fields.

From the results obtained, it was observed that the nature of the temperature dependence of the elastic coefficients K_φ strongly depends not only on the type of liquid crystal but much more on the nature of the phase transition to the adjacent phase. The electro-optic coefficient disappears rapidly (and, in practice, radically changes its value and character) in the case of the transition to the, also chiral, cholesteric N* phase. Moving to the paraelectric A* phase, the elastic coefficient K_φ begins to decrease rapidly approaching the phase transition and disappears at the temperature of the transition to the nonchiral phase. The situation is different for the antiferroelectric liquid crystal MHPOBC, in which the ferroelectric phase is bordered on both sides by chiral ferroelectric phases: SmC_α* and SmC_γ*. In this case, it can be assumed that K_φ will vanish monotonically only in the SmC_α* phase. Figure 6 shows the results of the temperature dependence of the coefficient of confinement Ka for chiral liquid crystals with phases with antiferroelectric properties D12 [66] and MHPOBC [71].

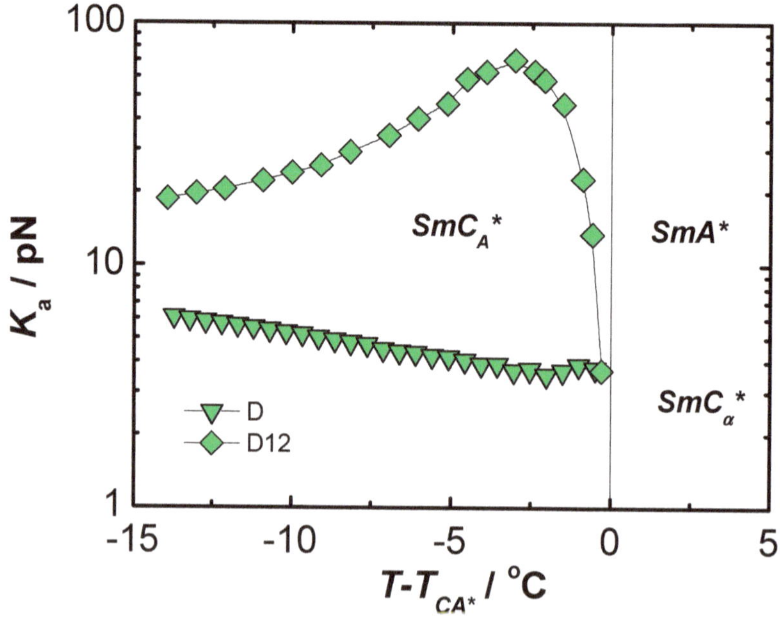

Figure 6. Temperature dependence of the elastic coefficient K_a in chiral liquid crystals with antiferroelectric phase: D12 (rhomboids) and MHPOBC (triangles) obtained by electro-optical method using low electric fields.

As shown in Figure 6, K_a exhibits a fairly strong temperature dependence, especially near the critical temperature of the transition from the SmC* to SmA* phase. In the paraelectric phase, the parameter K_a has a value of zero (by definition). It is also noteworthy that K_a takes relatively large values compared to the generally accepted magnitudes of the elastic constant in ferroelectric liquid crystals (Figure 3) [68]. This can be interpreted as a stiffening of the system due to the nearly antiparallel orientation of the electric moments of molecules in adjacent smectic layers. As the temperature increases below the critical temperature T_C, p increases and δP simultaneously decreases, leading to an increase in the time-averaged dipole–dipole interaction energy between molecules of adjacent smectic layers [66]. Consequently, as the temperature increases, the stiffening effect becomes stronger and, thus, K_a initially increases. The subsequent sharp decrease in the elastic coefficient K_a near $T = T_C$ is caused by the rapid disappearance of the tilt angle in the critical region. The elasticity constant in the MHPOBC material in the antiferroelectric phase gently decreases its value with increasing temperature.

The scatter of experimental results obtained for different methods and for different materials is quite large. For constant elasticity and viscosity, the experimental results can differ by orders of magnitude. The reasons for these discrepancies in helicoidal materials may be due to various aspects: on the one hand, nonlinear effects (the definitions of viscosity and elasticity coefficients assume linear torque dependence of angular velocity and strain, which is only true for laminar flows and small deformations) and, on the other hand, boundary conditions such as confining surfaces, anchorage, and gauge cell thickness (especially relevant for the direction of the screw, which, in thin planar cells, may be underdeveloped or deformed). An additional cause of discrepancies (as illustrated in Figure 7) turns out to be the influence of the sequence of phases, their chirality, and susceptibility to the electric field (the same phase in different materials shows different viscoelastic effects due to the proximity of a chiral or nonchiral, dielectric or paraelectric phase, etc.). Therefore, large deformations or turbulent flows during measurement can cause significant systematic errors. For this reason, viscoelastic properties should also be measured under defined, reproducible conditions, taking into account the helix pitch and its undisturbed distribution.

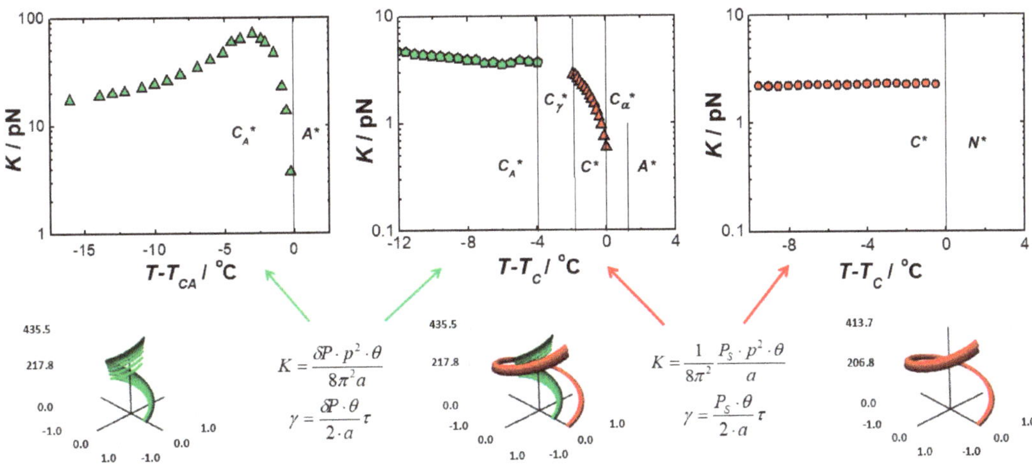

Figure 7. Elasticity constant K_φ as a function of temperature for three different materials in the low electric field range: D12, MHPOBC, and Ce 3 showing the different dependence of the elastic constant depending on the type of adjacent phases. To illustrate the difference in helix behavior, schematic drawings are shown for one pitch period for the compounds studied. Green indicates the antiferroelectric phase C_A*, red the ferroelectric phase C*.

In general, the determination of viscoelastic properties is not a simple task, as previously highlighted. The main problem is the multitude of physical parameters needed to determine the values of the elasticity and viscosity constants. It is also crucial to take into account the nature of the phase in question. This issue is particularly relevant for chiral phases, such as ferroelectric and antiferroelectric phases. There are several experimental methods for measuring viscosity and elasticity constants (especially viscosity) in chiral phases. These methods use various phenomena to detect deformation, such as light transmission, polarization current, light modulation, dielectric constant, and deformation or helix unfolding. Typically, an external electric field is used to induce deformation, and it is important that the electric field is uniform within the cell. Methods for testing viscoelastic properties are relatively well defined for liquid crystalline materials with nematic phases. Over the years, this methodology has been developed and extended both theoretically and experimentally. There are now new reports on, for example, chiral active nematicides (BPs) [78] or nonchiral nematics with ferroelectric properties [79], in which it has been shown that the polar orientation order of molecules can induce chirality in soft matter without chemically induced chiral centers, which consequently also affects their rheological properties. New technological solutions are already emerging, for example INSTEC's automatic liquid crystal testing (ALCT), which allows the automated determination of splay K_{11}, bend K_{33}, and twist K_{22} elastic constants in nematics with positive and negative dielectric anisotropy [80]. The situation with smectic phases is much more complicated. Quite a few theoretical models and experiments based on computer simulations are being developed. Simulations concern systems with nematic phases and smectic phases, with and without defects, with properties of hybrid systems with micro- and nanoparticles [81]. Theoretical models include assumptions to simplify calculations, for example, that smectic layers are incompressible (the angle of inclination θ remains constant), isotropies of elastic constants are assumed, or the absence of flexoelectricity and dielectric coupling. Methods for determining rotational viscosity using strong electric fields are being systematically extended, modified, and improved. There is also the possibility of automated measurement of FLC rotational viscosity proposed by INSTEC, using a specialized measurement attachment [80]. Our group focused on chiral ferroelectric and antiferroelectric liquid crystals in surface-stabilized cells. Sample homogeneity, a linear range of light modulation depth from the applied voltage, was assumed. Simultaneous calibration synchronized with the measurement of the electro-optical response provides independence from the influence of existing defects, dispersion, electric field effects, or ageing effects. To date, the selected materials have also not shown water absorption effects. In future work, it will be important to consider additional parameters that may affect the viscoelastic properties of chiral liquid crystals (temperature, humidity, homogeneity, ageing, doping, etc.). Uniformity and generalizability of conclusions is a priority for future research and solutions. The creation of a broad and accessible experimental database can be an indispensable source of information to optimize research and select the most effective methods for determining the elasticity and viscosity constants of liquid crystals [82]. Given the multitude of approaches, solutions, and questions, it is possible that machine learning will prove to be a helpful solution for the future.

5. Conclusions

Viscoelastic properties are one of the most fundamental properties of chiral liquid crystals. In general, their determination is not a simple task. The main problem is the multitude of physical parameters needed to determine the values of elasticity and viscosity constants. It is also necessary to take into account the nature of the phase in question. This problem is particularly important for chiral phases, such as ferroelectric and antiferroelectric phases. There are several experimental methods for measuring viscosity and elasticity constants in chiral phases. These methods use various phenomena to detect deformation, such as light transmission, polarization current, light modulation, dielectric constant, and helix deformation or unfolding. Usually, an external electric field is used to induce deformation, and it is essential that the electric field is uniform within the cell. The method of determining the coefficient of elasticity K_a or rotational viscosity γ is often determined by the purpose for which the material under study is to be used. If we need information on mechanical constants of the volumetric type, then the use of a thick homeotropic cell for all the parameters to be determined proves to be the most reliable. Liquid crystal materials with ferroelectric and antiferroelectric properties are most often used in light modulators, displays, or smart films, where the liquid crystal is surface-stabilized. In this case, a method using a linear electro-optical coefficient a to determine viscoelastic properties while maintaining the low deformation condition is most suitable.

Nevertheless, the main challenge should be a more critical and unambiguous analysis of the observed discrepancies. Based on the results and observations, the conclusion is that further research is needed that takes into account a larger number of materials, combinations of phases, and standardized experimental conditions that will allow the generalization of the obtained conclusions.

Funding: The work was supported by the Polish National Science Center, Project No. 2019/03/X/ST3/01516.

Data Availability Statement: The raw data supporting the conclusions of this article will be made available by the authors on request.

Acknowledgments: The work presented was inspired by the substantive part of the author's application for scientific promotion (habilitation thesis).

Conflicts of Interest: The author declares no conflicts of interest.

References

1. Lagerwall, S.T. Chirality, Symmetry and Physical Effects. In *Chiral Liquid Crystals*; Kuczyński, W., Ed.; IFM PAN: Poznań, Poland, 2005; pp. 185–222.
2. Kitzerow, H.; Bahr, C. *Chirality in Liquid Crystals*; Springer Science & Business Media: Berlin/Heidelberg, Germany, 2001.
3. Sluckin, T.J.; Dunmur, D.A.; Stegemeyer, H. *Crystals That Flow: Classic Papers from the History of Liquid Crystals*; Taylor & Francis: Abingdon, UK, 2004.
4. Srivastava, A.K.; Chigrinov, V.G.; Kwok, H.S. Ferroelectric liquid crystals: Excellent tool for modern displays and photonics. *J. Soc. Inf. Disp.* **2015**, *23*, 253–272. [CrossRef]
5. Pontus, J.; Dahlgren, A.; Komitov, L.; Matharu, A.S.; Grover, C.; Gouda, F.; Kutub, A. Optic, electrooptic and dielectric properties of novel antiferroelectric liquid crystal compounds. *Ferroelectrics* **2000**, *244*, 147–157. [CrossRef]
6. Guo, Q.; Kexin, Y.; Chigrinov, V.; Zhao, H.; Tribelsky, M. Ferroelectric Liquid Crystals: Physics and Applications. *Crystal* **2019**, *9*, 470. [CrossRef]
7. Chandani, A.D.I.; Górecka, E.; Ouchi, Y.; Takezoe, H.; Fukuda, A. Antiferroelectric Chiral Smectic Phases Responsible for the Trislable Switching in MHPOBC. *Jpn. J. Appl. Phys.* **1998**, *28*, L1265–L1268. [CrossRef]
8. Fitas, J.; Marzec, M.; Szymkowiak, M.; Jaworska-Gołąb, T.; Deptuch, A.; Tykarska, M.; Kurp, K.; Urbańska, M.; Bubnov, A. Mesomorphic, electro-optic and structural properties of binary liquid crystalline mixtures with ferroelectric and antiferroelectric liquid crystalline behavior. *Phase Transit.* **2018**, *91*, 1017–1026. [CrossRef]
9. Kuczyński, W. Electro-Optical Studies of Relaxation Processes in Chiral Smectic Liquid Crystals. In *Relaxation Phenomena: Liquid Crystals, Magnetic Systems, Polymers, High-Tc Superconductors, Metalic Glasses*; Haase, W., Wróbel, S., Eds.; Springer: Berlin/Heidelberg, Germany, 2003; pp. 422–444.
10. Orihara, H.; Ishibashi, Y. Electro-Optic Effect and Third-Order Nonlinear Dielectric Response in Antiferroelectric Liquid Crystals. *J. Phys. Soc. Jpn.* **1995**, *64*, 3775–3786. [CrossRef]
11. Dardas, D.; Kuczyński, W. Non-linear electrooptical effects in chiral liquid crystals. *Opto-Electron. Rev.* **2004**, *12*, 277–280.
12. Dardas, D.; Kuczyński, W.; Hoffmann, J. Measurements of absolute values of electrooptic coefficients in a ferroelectric liquid crystal. *Phase Transit.* **2006**, *79*, 213–222. [CrossRef]
13. Jeżewski, W.; Kuczyński, W.; Hoffmann, J.; Dardas, D. Comparison of dielectric and optical response of chevron ferroelectric liquid crystals. *Opto-Electron. Rev.* **2008**, *16*, 281–286. [CrossRef]
14. Kuczyński, W.; Dardas, D.; Goc, F.; Dabrowski, R. Linear and quadratic electrooptic effects in antiferroelectric liquid crystals. *Ferroelectrics* **2000**, *244*, 191–199. [CrossRef]
15. Lagerwall, S.T. Linear and Non-Linear Polar Effects in Liquid Crystals. *Mol. Cryst. Liq. Cryst.* **2011**, *543*, 769–813. [CrossRef]
16. Buka, A.; Éber, N. (Eds.) *Flexoelectricity in Liquid Crystals: Theory, Experiments and Applications*; World Scientific: Singapore, 2012.
17. Gurovich, E.V.; Kats, E.I.; Lebedev, V.V. Orientational phase transitions in liquid crystals. Critical dynamics. *Zh. Eksp. Teor. Fiz* **1991**, *100*, 891.
18. Stewart, I.W. *The Static and Dynamic Continuum Theory of Liquid Crystals*; CRC Press: Boca Raton, FL, USA, 2004.
19. Jakli, A.; Saupe, A. *One- and Two Dimensional Fluids: Properties of Smectic, Lamellar and Columnar Liquid Crystals*; CRC Press: Boca Raton, FL, USA, 2006.
20. Sengupta, A. Liquid Crystal Theory. In *Topological Microfluidics*; Springer Theses; Springer: Berlin/Heidelberg, Germany, 2013.
21. de Genes, P.G. *The Physics of Liquid Crystals*; Oxford University Press: Oxford, UK, 1974.
22. Męsowicz, M. The three coefficients of viscosity of anisotropic liquids. *Nature* **1946**, *158*, 27. [CrossRef] [PubMed]
23. Gähwiller, C. Direct determination of the five independent viscosity coefficients of nematic liquid crystals. *Mol. Cryst. Liq. Cryst.* **1973**, *20*, 301–318. [CrossRef]
24. Leslie, F.M.; Laverty, J.S.; Carlsson, T. Continuum theory for biaxial nematic liquid crystals. *Q. J. Mech. Appl. Math.* **1992**, *45*, 595–606. [CrossRef]
25. Leslie, F.M.; Stewart, I.W.; Nakagawa, M. A continuum theory for smectic C liquid crystals. *Mol. Cryst. Liq. Cryst.* **1991**, *198*, 443–454. [CrossRef]
26. Takezoe, H.; Kondo, K.; Miyasato, K.; Abe, S.; Tsuchiya, T.; Fukuda, A.; Kuze, E. On the methods of determining material constants in ferroelectric smectic C* liquid crystals. *Ferroelectrics* **1984**, *58*, 55. [CrossRef]
27. Lagerwall, S.T. *Ferroelectric and Antiferroelectric Liquid Crystals*; Wiley-VCH: Weinheim, Germany, 1999.
28. Lagerwall, J.P.F.; Giesselmann, F. Current topics in smectic liquid crystal research. *Chem. Phys. Chem.* **2006**, *7*, 20–45. [CrossRef] [PubMed]
29. Dahl, I.; Lagerwall, S.T. Elastic and flexoelectric properties of chiral smectic-C phase and symmetry considerations on ferroelectric liquid-crystal cells. *Ferroelectrics* **1984**, *58*, 215. [CrossRef]
30. Skarp, K. Rotational viscosities in ferroelectric smectic liquid crystals. *Ferroelectrics* **1988**, *84*, 119. [CrossRef]
31. Larson, R.G. *The Structure and Rheology of Complex Fluids*; Oxford University Press: New York, USA, 1999.
32. Orsay Liquid Crystal Group. Simplified elastic theory for smectics C. *Solid State Commun.* **1971**, *9*, 653. [CrossRef]
33. Ericksen, J.L. Anisotropic fluids. *Arch. Ration. Mech. Anal.* **1960**, *4*, 231–237. [CrossRef]
34. Parodi, O. Stress tensor for a nematic liquid crystal. *J. Phys.* **1970**, *31*, 581–584. [CrossRef]
35. Handschy, M.A.; Clark, N.A. Structures and responses of ferroelectric liquid crystals in the surface-stabilized geometry. *Ferroelectrics* **1984**, *59*, 69–116. [CrossRef]

36. Escher, C.; Geelhaar, T.; Böhm, E. Measurement of the rotational viscosity of ferroelectric liquid crystals based on a simple dynamical model. *Liq. Cryst.* **1988**, *3*, 469–484. [CrossRef]
37. Piecek, W.; Rutkowska, J.; Kedzierski, J.; Perkowski, P.; Raszewski, Z.; Dabrowski, R.R. *Liquid Crystals: Physics, Technology and Applications*; SPIE: Bellingham, WA, USA, 1997; Volume 3318, pp. 148–153.
38. Pierański, P.; Guyon, E.; Keller, P. Shear flow induced polarization in ferroelectric smectics C. *J. Phys.* **1975**, *36*, 1005. [CrossRef]
39. Clark, N.A.; Lagerwali, S.T. Submicrosecond bistable electro-optic switching in liquid crystals. *Appl. Phys. Lett.* **1980**, *36*, 899. [CrossRef]
40. Flatischler, K.; Skarp, K.; Lagerwall, S.T.; Stebler, B. Thin ferroelectric liquid crystal cells with surface-induced alignment: Optical characterization and electro-optic performance. *Mol. Cryst. Liq. Cryst.* **1985**, *131*, 21. [CrossRef]
41. Dahl, I.; Lagerwall, S.T.; Skarp, K. Simple model for the polarization reversal current in a ferroelectric liquid crystal. *Phys. Rev. A* **1988**, *36*, 4380. [CrossRef]
42. Geelhaar, T. Ferroelectric mixtures and their physico-chemical properties. *Ferroelectrics* **1988**, *85*, 329–349. [CrossRef]
43. Kuczyński, W. Determination of elasticity and viscosity coefficients in a ferroelectric smectic c liquid crystal. *Ber. Bunseeges. Phys. Chem.* **1981**, *85*, 234. [CrossRef]
44. Kuczyński, W.; Lagerwall, S.T.; Stebler, B. Determination of Elasticity and Viscosity Coefficients in a Chiral Smectic C Liquid Crystal. In Proceedings of the 11th Internationa Liquid Crystal Conference, Berkeley, CA, USA, 30 June–4 July 1986.
45. Kawaida, M.; Akahane, T. Measurement of elastic constant and intrinsic twist of ferroelectric SmC* liquid crystal. *Jap. J. Appl. Phys.* **1990**, *29*, 340. [CrossRef]
46. Ezcurra, A.; Jubindo, M.A.P.; De La Fuente, M.R.; Etxebarria, J.; Remon, A.; Tello, M.J. Dielectric and optical studies near the chiral-smectic-C–smectic-A transition of (s)-2-hydroxy-4-decyloxybenzylidene-4′-amino-2p″-methylbutylcinnamate. *Liq. Cryst.* **1989**, *4*, 125. [CrossRef]
47. Gouda, F.; Skarp, K.; Andersson, G.; Kresse, H.; Lagerwall, S.T. Viscoelastic properties of the smectic A* and C* phases studied by a new dielectric method. *Jap. J. Appl. Phys.* **1989**, *28*, 1887. [CrossRef]
48. Levstik, A.; Kutnjak, Z.; Levstik, J.; Zeks, B. Rotational viscosity in ferroelectric liquid crystal dobambc determined by dielectric method. *Ferroelectrics* **1990**, *109*, 267. [CrossRef]
49. Adamow, A.; Sznitko, L.; Chrzumnicka, E.; Stachera, J.; Szukalski, A.; Martyński, T.; Myśliwiec, J. The ultra-photostable and electrically modulated Stimulated Emission in perylene-based dye doped liquid crystal. *Sci. Rep.* **2019**, *9*, 2143. [CrossRef] [PubMed]
50. Shen, Y.; Dierking, I. Perspectives in Liquid-Crystal-Aided Nanotechnology and Nanoscience. *Appl. Sci* **2019**, *9*, 2512. [CrossRef]
51. Hamley, I.W. Liquid Crystal Phase Formation by Biopolymers. *Soft Matter* **2010**, *6*, 1863–1871. [CrossRef]
52. Tian, L.; Goodby, J.W.; Görtz, V.; Gleeson, H.F. The magnitude and temperature dependence of the Kerr constant in liquid crystal blue phases and the dark conglomerate phase. *Liq. Cryst.* **2013**, *40*, 1446. [CrossRef]
53. Knapkiewicz, M.; Rachocki, A.; Bielejewski, M.; Sebastiao, P.J. NMR studies of molecular ordering and molecular dynamics in a chiral liquid crystal with the SmC_{α}* phase. *Phys. Rev. E* **2020**, *101*, 52708. [CrossRef]
54. Liang, H.H.; Xiao, Y.Z.; Hsh, F.J.; Wu, C.C.; Lee, J.Y. Enhancing the electro-optical properties of ferroelectric liquid crystals by doping ferroelectric nanoparticles. *Liq. Cryst.* **2010**, *37*, 255–261. [CrossRef]
55. Ghosh, S.; Roy, S.K.; Acharya, S.; Chakrabarti, P.K.; Zurowska, M.; Dabrowski, R. Effect of multiferroic $BiFeO_3$ nanoparticles on electro-optical and dielectric properties of a partially fluorinated orthoconic antiferroelectric liquid crystal mixture. *EPL* **2011**, *96*, 47003. [CrossRef]
56. Resetic, A.; Milavec, J.; Domenici, V.; Zupancic, B.; Bubnov, A.; Zalar, B. Deuteron NMR investigation on orientational order parameter in polymer dispersed liquid crystal elastomers. *Phys. Chem. Chem. Phys.* **2020**, *22*, 23064–23072. [CrossRef] [PubMed]
57. Khan, R.K.; Turlapati, S.; Rao, N.V.S.; Ghosh, S. Elastic and dielectric properties of ferroelectric nanoparticles/bent-core nematic liquid crystal blend. *Eur. Phys. J. E* **2017**, *40*, 75. [CrossRef] [PubMed]
58. Singh, A.K.; Iqbal, A.; Singh, U.B.; Dabrowski, R.S.; Dhar, R. Thermodynamic, optical and switching parameters of a ferroelectric liquid crystalline material having SmA*-SmC*-SmBh* phase sequence. *Phase Transit.* **2018**, *91*, 811–820. [CrossRef]
59. Zgueb, R.; Dhaouadi, H. Theoretical and experimental analysis of surface anchoring in the surface stabilization of ferroelectric liquid crystal cells. *Capillarity* **2024**, *11*, 31–40. [CrossRef]
60. Lavrentovich, O.D. Splay-bend elastic inequalities shape tactoids, toroids, umbilics, and conic section walls in paraelectric, twist-bend, and ferroelectric nematics. *Liq. Cryst. Rev.* **2024**, *12*, 1–13. [CrossRef]
61. Pierański, P.; Guyon, E.; Keller, P.; Liebert, L.; Kuczyński, W.; Pierański, P. Optical Study of a Chiral Smectic C Under Shear. *Mol. Cryst. Liq. Cryst.* **1977**, *38*, 275–301. [CrossRef]
62. Kuczyński, W.; Nowicka, K.; Dardas, D.; Jeżewski, W.; Hoffmann, J. Determination of bulk values of twist elasticity coefficient in a chiral smectic C* liquid crystal. *Opto-Electron. Rev.* **2010**, *18*, 176–180. [CrossRef]
63. Kuczyński, W.; Dardas, D.; Nowicka, K. Determination of the bulk rotational viscosity coefficient in a chiral smectic C* liquid crystal. *Phase Transit.* **2009**, *82*, 444–451. [CrossRef]
64. Panarin, Y.P.; Kalinovskaya, O.; Vij, J.K. The investigation of the relaxation processes in antiferroelectric liquid crystals by broad band dielectric and electro-optic spectroscopy. *Liq. Cryst.* **1998**, *25*, 241–252. [CrossRef]
65. Dardas, D.; Kuczyński, W.; Hoffmann, J.; Jeżewski, W.; Nowicka, K.; Małecki, J. Non-linear electrooptic effect in antiferroelectric liquid crystal. *Opto-Electron. Rev.* **2009**, *17*, 25–29. [CrossRef]

66. Dardas, D.; Kuczyński, W.; Hoffmann, J.; Jeżewski, W. Determination of twist elastic constant in antiferroelectric liquid crystals. *Meas. Sci. Technol.* **2011**, *22*, 085707. [CrossRef]
67. Diamant, H.; Drenck, K.; Pepinsky, R. Bridge for accurate measurement of ferroelectric hysteresis. *Rev. Sci. Instrum.* **1957**, *28*, 30. [CrossRef]
68. Kuczyński, W.; Dardas, D.; Hoffmann, J.; Nowicka, K.; Jeżewski, W. Comparison of methods for determination of viscoelastic properties in chiral smectics C*. *Phase Transit.* **2012**, *85*, 358–363. [CrossRef]
69. Dardas, D. Electro-optic and viscoelastic properties of a ferroelectric liquid crystalline binary mixture. *Phase Transit.* **2016**, *89*, 368–375. [CrossRef]
70. Dardas, D. Tuning the electro-optic and viscoelastic properties of ferroelectric liquid crystalline materials. *Rheol. Acta* **2019**, *58*, 193–201. [CrossRef]
71. Dardas, D.; Lalik, S.; Nowacka, Z.; Yevchenko, T.; Marzec, M. Electro-Optic Effect of Laser Photobleaching on Viscoelastic Properties of Chiral Liquid Crystals. *Crystals* **2023**, *13*, 164. [CrossRef]
72. Yang, F.; Bradberry, G.W.; Sambles, J.R. Optical determination of the twist elastic constant of a smectic-C* liquid crystal. *Phys. Rev. E* **1996**, *53*, 674. [CrossRef]
73. Rosenblatt, C.; Pindak, R.; Clark, N.A.; Meyer, R.B. Freely Suspended Ferroelectric Liquid-Crystal Films: Absolute Measurements of Polarization, Elastic Constants, and Viscosities. *Phys. Rev. Lett.* **1979**, *42*, 1220. [CrossRef]
74. Sah, Y.; Suresh, K.A. Visco-elastic modes in some ferro-electric liquid crystals. *Liq. Cryst.* **1998**, *24*, 701. [CrossRef]
75. Kuczyński, W. Behavior of the helix in some chiral smectic-C* liquid crystals. *Phys. Rev. E* **2010**, *81*, 021708. [CrossRef] [PubMed]
76. Meyer, R.B. Ferroelectric Liquid Crystals; A Review. *Mol. Cryst. Liq. Cryst.* **1977**, *40*, 33. [CrossRef]
77. Kumari, P.; Basnet, B.; Lavrentovich, M.O.; Lavrentovich, O.D. Chiral ground states of ferroelectric liquid crystals. *Science* **2024**, *383*, 1364–1368. [CrossRef] [PubMed]
78. Stannarius, R. Liquid crystals: More than display fillings. *Nat. Mater.* **2009**, *8*, 617–618. [CrossRef] [PubMed]
79. Kralj, N.; Ravnik, M.; Kos, Ž. Chirality, anisotropic viscosity and elastic anisotropy in three-dimensional active nematic turbulence. *Commun. Phys.* **2024**, *7*, 222. [CrossRef]
80. INSTEC Inc. Available online: https://instec.com/portal/list/index/id/13.html (accessed on 26 May 2024).
81. Emelyanenko, A.V.; Filimonova, E.S.; Khokhlov, A.R. Molecular origin of the heterogeneity in the nematic and smectic liquid crystals: Elastic constants, gradients of order parameters, and visualization of small objects. *Phys. Rev. E* **2021**, *103*, 022709. [CrossRef]
82. Osiecka-Drewniak, N.; Drzewicz, A.; Juszyńska-Gałązka, E. Machine learning studies for liquid crystal texture recognition. *Liq. Cryst.* **2023**, *51*, 255–264. [CrossRef]

MDPI AG
Grosspeteranlage 5
4052 Basel
Switzerland
Tel.: +41 61 683 77 34

Materials Editorial Office
E-mail: materials@mdpi.com
www.mdpi.com/journal/materials

www.ingramcontent.com/pod-product-compliance
Lightning Source LLC
Chambersburg PA
CBHW062040220926
51850CB00010B/2616
9783725867622